Systematic Materials Analysis

VOLUME II

MATERIALS SCIENCE AND TECHNOLOGY

A. S. Nowick and B. S. Berry, ANELASTIC RELAXATION IN CRYSTALLINE SOLIDS, 1972

E. A. Nesbitt and J. H. Wernick, RARE EARTH PERMANENT MAGNETS, 1973

W. E. Wallace, RARE EARTH INTERMETALLICS, 1973

J. C. Phillips, BONDS AND BANDS IN SEMICONDUCTORS, 1973

H. Schmalzried, SOLID STATE REACTIONS, 1974

J. H. Richardson and R. V. Peterson (editors), SYSTEMATIC MATERIALS ANALYSIS, VOLUMES I and II, 1974. Volume III in preparation

Systematic Materials Analysis

VOLUME II

Edited by

J. H. RICHARDSON

Materials Sciences Laboratory
The Aerospace Corporation
El Segundo, California

R. V. PETERSON

Materials Sciences Laboratory
The Aerospace Corporation
El Segundo, California

ACADEMIC PRESS New York and London 1974

A Subsidiary of Harcourt Brace Jovanovich, Publishers

ACADEMIC PRESS, INC.
111 Fifth Avenue, New York, New York 10003

United Kingdom Edition published by
ACADEMIC PRESS, INC. (LONDON) LTD.
24/28 Oval Road, London NW1

Library of Congress Cataloging in Publication Data

Richardson, James H
Systematic materials analysis.

(Materials science series)
Includes bibliographies.
1. Materials–Analysis. 2. Instrumental analysis.
I. Peterson, Ronald V., joint author. II. Title.
QD131.R48 543 72-12203
ISBN 0–12–587802–8 (v. 2)

PRINTED IN THE UNITED STATES OF AMERICA

Dedicated to the One Perfect Instrument:

HEBREWS 1:2
JOHN 3:17

J.H.R.
R.V.P.

Contents

Chapter 14 Raman Spectrometry

J. E. Katon

Chapter 15 Refractometry

J. H. Richardson

Chapter 16 Scanning Electron Microscopy

John C. Russ

Chapter 17 Ultraviolet Photoelectron Spectrometry

John J. Uebbing

Chapter 18 Visible and Ultraviolet Absorption Spectrometry

Richard S. Danchik

Chapter 19 X-Ray Photoelectron Spectrometry (ESCA)

Warren G. Proctor

List of Contributors

Numbers in parentheses indicate the pages on which the authors' contributions begin.

D. BRYAN (45), Gulf Radiation Technology, San Diego, California

R. S. CODRINGTON (73), Varian Associates, Palo Alto, California

ROBERT L. COOK (1), Mississippi State University, Mississippi State, Mississippi

RICHARD S. DANCHIK (199), Aluminum Company of America, Alcoa Technical Center, Alcoa Center, Pennsylvania

GORDON E. JONES (1), Mississippi State University, Mississippi State, Mississippi

J. E. KATON (119), Miami University, Oxford, Ohio

H. R. LUKENS* (45), Gulf Radiation Technology, San Diego, California

WARREN G. PROCTOR (229), Varian Associates, Palo Alto, California

J. H. RICHARDSON (143), The Aerospace Corporation, El Segundo, California

JOHN C. RUSS (159), EDAX Laboratories, Raleigh, North Carolina

H. L. SCHLESINGER (45), Gulf Radiation Technology, San Diego, California

JOHN J. UEBBING† (183), Varian Associates, Palo Alto, California

DON WARE‡ (73), Varian Associates, Palo Alto, California

* Present address: Intelcom Rad Tech, San Diego, California.
† Present address: Hewlett-Packard, Palo Alto, California.
‡ Present address: Bruker Magnetics, Inc., Burlington, Massachusetts.

Preface

It is both exciting and dismaying to observe the parade of new and refined instrumental methods available for the analysis of materials—exciting because these instruments provide opportunities for faster and more reliable answers to material analysis problems, dismaying because one is hard pressed to evaluate these various instruments for a given task. Materials analysis often involves the complete characterization of a material, including structural and textural analyses in addition to chemical analysis.

It has been the aim of the editors of *Systematic Materials Analysis* to satisfy the needs of the materials analyst in these areas by presenting brief discussions on a broad range of instrumental methods and bringing to their selection new approaches that will yield the desired information about a given material. These volumes not only comprise a brief, comprehensive reference for the materials analyst but also provide a source of information for the engineer or researcher who must select the appropriate instrument for his immediate needs. Although the volumes are directed toward the physical sciences, they can also be of value for the biological scientist with materials problems and of use to the laboratory administrator as both convenient reference and guide for the purchase of new instrumentation.

Chapter 1 focuses on the selection of analytical methods on the bases of specimen limitations and information desired. The selection is made by use of flow charts encompassing the various instruments outlined in the succeeding chapters. The unique character and utility of this work lie in the use of these charts, since they present a complete listing of analytical instrumentation arranged so as to permit selection of the best method(s) for a given analytical task. The student may thus gain insights into thought processes that are usually acquired only after years of experience in this field. Thus, these volumes can appropriately serve as a college text (third year to graduate level) as well as a reference work.

The chapters on specific instruments briefly outline the theories of operation, with detailed discussions of theory fully referenced, and describe the capability of the methods for qualitative and quantitative measurements of chemical composition, structure, and texture (as applicable).

Topics such as the sensitivity and selectivity of each method are emphasized. References illustrating the operation of the instrument, as well as references to user-constructed accessories that extend and improve the instrument's capabilities, are included when applicable.

The wide variety of commercial instruments available precludes the inclusion of instructions for the operation of instruments and, consequently, the inclusion for the student of experiments based on these instructions. For the same reason, comprehensive descriptions and the inevitable comparisons of commercial instruments are beyond the scope of this work.

Acknowledgments

We want to thank all the authors of this work for their willing participation in this endeavor, and we gratefully acknowledge their corrections and comments on the flow charts in Chapter 1.

We also want to thank our many colleagues at The Aerospace Corporation who gave support in various ways, especially Mrs. Genevieve Denault, Camille Gaulin, Dr. Wendell Graven, Henry Judeikis, Dr. Gary Stupian, and Dr. Hideyo Takimoto, who rendered specific suggestions and reviewed chapters. We remember with special affection the late Dr. Thomas Lee, whose remarks and comments were very valuable in the development of the concept of this work.

We also wish to thank Miss Debra Levy and Mrs. Myra Peterson for help in the critical review of the work and Miss Rosalie Hernandez, Mrs. Jean Hill, Mrs. Carolyn Thompson, and Mrs. Marsha Graven for typing assistance.

To Ann and Myra Ann, our wives, we are grateful for their love and their spiritual challenge to us.

Contents of Other Volumes

Volume I

Volume III

Volume IV (tentative)

CHAPTER 11

Microwave Spectrometry

Robert L. Cook and Gordon E. Jones

Mississippi State University
Mississippi State, Mississippi

Introduction

Microwave rotational spectroscopy deals with the absorption of electromagnetic radiation in the region between conventional radio waves and infrared waves. The absorption spectra arise from the molecular rotation and correspond to transitions between rotational energy levels. These transitions are induced through the interaction of the molecular electric dipole with the electric vector of the radiation field. Molecules having microwave rotational spectra are essentially limited to polar gases. Weak magnetic dipole transitions are also possible; however, few stable molecules have a magnetic dipole. Roughly, the microwave region extends from around 1000 MHz (30 cm) to 1,000,000 MHz (0.3 mm). In this spectral region, frequencies are expressed in megahertz (MHz) or gigahertz (GHz), where the hertz (Hz) unit denotes cycles per second, and 1 MHz $= 10^6$ Hz, 1 GHz $= 10^9$ Hz. In terms of wavelengths, 30 GHz and 300 GHz correspond to 1 cm and 1 mm, respectively. The highest frequency so far detected (Helminger *et al.*, 1970) for a rotational transition in the submillimeter wave region (frequencies beyond 300 GHz) is 813 GHz (0.368 mm).

The field of microwave rotational spectroscopy has grown rapidly since the development of the various radar components during World War II, and an enormous amount of data on molecular properties has been obtained. The information which can be obtained from a microwave study includes precise molecular structures, dipole moments, centrifugal distortion constants, vibrational potential functions, internal rotation barriers, nuclear masses and spins, nuclear quadrupole coupling constants, molecular magnetic moments, conformations of rotational isomers, magnetic susceptibility constants, molecular quadrupole moments, and much more. A recent compilation of molecular structures obtained from microwave spectroscopy is given by Gordy and Cook (1970).

Early in the development of microwave spectroscopy it was realized that the rotational spectrum could be employed for chemical analysis. Each molecule has its own characteristic spectrum, and the location of the absorption lines provides the basis for a qualitative analysis. The intensities of the absorption lines provide the basis for a quantitative analysis. Recent reviews of the subject have been given by Lide (1966), Scharpen and Laurie (1972), and Sheridan (1973). Although microwave spectroscopy has been a powerful research tool for over 25 years, it has not yet matured to its full potential as an analytical technique. The cataloging of spectra has not yet advanced to the necessary stage of development, and certain practical problems of its application need further study. Nevertheless, there are numerous problems for which microwave spectroscopy will provide a unique and simple analytical method. The development of a commercial microwave spectrometer, along with recent theoretical developments, and the interest of NASA in the use of microwave spectroscopy as a means of detecting atmospheric contaminants have provided significant impetus to its further development.

Since the emphasis will be on chemical analysis, only a brief introduction to the theory of rotational spectra can be given here. A number of books devoted to the subject of microwave spectroscopy are available, and the reader is directed to these for a more detailed discussion. See, for example, Gordy *et al.* (1953), Townes and Schawlow (1955), Sugden and Kenney (1965), Wollrab (1967), and Gordy and Cook (1970).

1 Theory of the Method

1.1 Rotational Energy Levels and Spectra

With each electronic state of a molecule, we may associate a set of vibrational energy levels, and with each vibrational state we have a set of rotational energy levels. In the study of pure rotational spectra, we are concerned with transitions between the various rotational sublevels associated with a given vibrational state of a particular electronic state. In general, the transitions studied are in the ground electronic state. On the other hand, rotational transitions in both the ground and excited vibrational states are observed.

To discuss the rotational problem, a molecule-fixed axis system is chosen whose origin coincides with the center of mass of the molecule and whose axes are oriented along the three mutually perpendicular principal inertial axes. The principal axes of inertia are designated by a, b, and c. The moments of inertia about the various principal axes are designated as I_a, I_b, and I_c where, by convention, the inertial axes a, b, c are so labeled that $I_a \leq I_b \leq I_c$. The principal types of rotors of interest to microwave spectroscopists may be classified according to the relative values of the principal moments of inertia. The three cases may be summarized as shown in Table 1.

Since molecular systems obey the laws of quantum mechanics, only certain well-defined rotational energies are allowed. According to quantum mechanics, both the total angular momentum, $P = \hbar[J(J + 1)]^{1/2}$, and its projection, $P_a = \hbar K$, along the symmetry axis are quantized for a symmetric top. The quantum number J can take on the values 0, 1, 2,

TABLE 1

TYPES OF MOLECULAR ROTORS[a]

Linear molecules	$I_a = 0,\ I_b = I_c$	Axis a along the internuclear axis, b and c perpendicular to this axis, e.g., OCS
Symmetric-top molecules		
Prolate top	$I_a < I_b = I_c$	Axis of least moment of inertia, a, along the symmetry axis, e.g., CH_3F
Oblate top	$I_a = I_b < I_c$	Axis of largest moment of inertia, c, along the symmetry axis, e.g., BCl_3
Asymmetric-top molecules	$I_a \neq I_b \neq I_c$	All three moments of inertia different, e.g., SO_2

[a] For a symmetric top, the molecule is designated a prolate or oblate rotor, depending on which inertia axis corresponds to the molecular symmetry axis. Most molecules belong to the asymmetric rotor case.

3, . . . , while K can have the values 0, ± 1, ± 2, . . . , $\pm J$. The energy levels for a prolate rotor are given by

$$E_{J,K}/h = BJ(J + 1) + (A - B)K^2 - D_J J^2(J + 1)^2 - D_{JK}J(J + 1)K^2 - D_K K^4 \qquad (1)$$

where $A = h/8\pi^2 I_a$, $B = h/8\pi^2 I_b$ are the rotational constants. For an oblate symmetric top, the unique axis is designated c, and the energy expression may be obtained from the above expression by replacement of A by C. The centrifugal distortion terms D_J, etc., account for the fact that a real molecule is not a rigid rotor. These constants are very small compared with the rotational constants, and their effects are important only for high J. It follows from Eq. (1) that the energy levels for $K \neq 0$ are doubly degenerate since the energy does not depend on the sign of K. There are then $(J + 1)$ different rotational sublevels for each J value.

The type of transitions which can occur between rotational energy levels is governed by quantum mechanical selection rules. For a transition to be allowed, the electric dipole moment matrix element (or transition moment), $\mu_{ij} = \int \Psi_i^* \boldsymbol{\mu} \Psi_j \, d\tau$, must be nonvanishing. For rotational absorption of radiation, the important selection rules for a symmetric top are $J \rightarrow J + 1$, $K \rightarrow K$, and the rotational frequencies are given by

$$\nu = (E_{J+1,K} - E_{J,K})/h = 2B(J + 1) - 4D_J(J + 1)^3 - 2D_{JK}(J + 1)K^2 \qquad (2)$$

Equations (1) and (2) also apply to a linear molecule for which $K = 0$. In the absence of centrifugal distortion effects, linear and symmetric-top molecules have similar spectra. The rigid rotor spectrum consists of a series of lines, $\nu = 2B, 4B, 6B, \ldots$, separated by $2B$. The presence of the distortion term D_{JK} for a symmetric top leads to the splitting of a given $J \rightarrow J + 1$ transition into $(J + 1)$ lines. The separation of these lines increases as K^2. Figure 1 shows schematically the spectral pattern of linear and symmetric-top molecules.

The evaluation of the rotational energy levels for an asymmetric rotor is more difficult than for a linear or symmetric top molecule. It is found that the rotational energies of a rigid asymmetric rotor can be written in the form

$$E_{J,\tau}/h = [(A + C)/2]J(J + 1) + [(A - C)/2]E_{J\tau}(\kappa) \qquad (3)$$

where $A > B > C$ and $E_{J\tau}(\kappa)$ is Ray's reduced energy. The parameter $\kappa = (2B - A - C)/(A - C)$ is a measure of the asymmetry. For a prolate symmetric top $B = C$ and $\kappa = -1$, while for an oblate top $A = B$ and

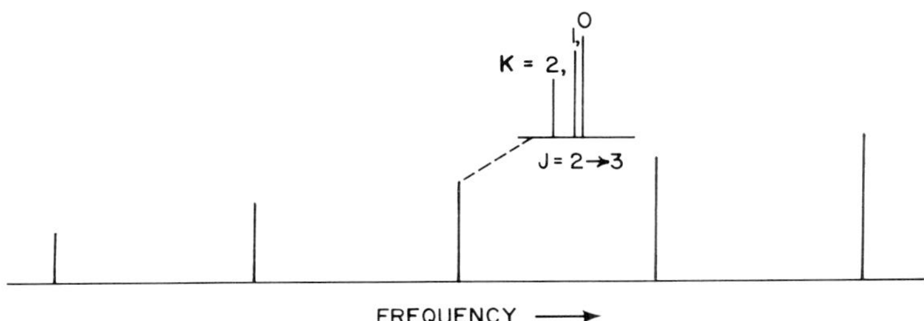

FIG. 1. General pattern of the rotational spectrum of a linear or symmetric-top molecule. The inset shows, for a symmetric top, the separation of the lines of different $|K|$ for the $J = 2 \rightarrow 3$ transition due to centrifugal distortion.

$\kappa = +1$. The range of κ is therefore $-1 \leq \kappa \leq 1$. The reduced energy depends on κ and the level in question, labeled by J and τ. The reduced energies have been tabulated for various asymmetries (Townes and Schawlow, 1955), and these tables may be used to evaluate the energy levels and the absorption frequencies. For each value of J there are $(2J + 1)$ different rotational sublevels. Although J is still a good quantum number, the quantum number K, useful for characterizing the symmetric top levels, is no longer a good quantum number. To distinguish the various levels of an asymmetric rotor, the notation J_τ is employed. The index τ, a pseudo-quantum number, is an integer ranging from $-J$ to $+J$. In this notation, the $(2J + 1)$ values of τ for a given J are assigned to the energy levels in the following order: The lowest energy sublevel is denoted by J_{-J}, the next lowest by J_{-J+1}, and so on to the highest level, which is denoted by J_{+J}. The three $J = 1$ levels in this notation are 1_{-1}, 1_0, 1_{+1}. An alternate notation which is more often employed is J_{K_{-1},K_1}. The subscript K_{-1} is the $|K|$ which characterizes the level in the prolate rotor limit $\kappa = -1$, and K_1 is the corresponding $|K|$ of the limiting oblate rotor $\kappa = 1$. In this notation the $J = 1$ levels are denoted by $1_{0,1}$, $1_{1,1}$, $1_{1,0}$. The relation between the two notations is $\tau = K_{-1} - K_1$.

For dipole absorption of radiation, the selection rules for J are $J \rightarrow J + 1$, $J \rightarrow J$, and $J \rightarrow J - 1$. These are designated R, Q, and P branch transitions, respectively. The selection rules for the subscripts K_{-1} and K_1 may be conveniently expressed in terms of the evenness (e) and oddness (o) of K_{-1}, K_1. In general, an asymmetric top can have dipole moment components along all three principal axes, denoted μ_a, μ_b, and μ_c. Transitions which arise from the μ_a component are designated as "a"-type transitions, and the allowed changes of K_{-1}, K_1 are ee $\leftrightarrow$ eo, oe $\leftrightarrow$ oo. For "b"-type transitions, the selection rules are ee $\leftrightarrow$ oo, oe $\leftrightarrow$ eo, and for the "c"-type transitions we have ee ↔ oe, eo ↔ oo. If a molecule lacks a particular dipole component, the corresponding transitions are not allowed. As an example, consider an asymmetric rotor with the dipole moment lying entirely along the b axis. Only b-type transitions are allowed, and the possible Q-branch transitions for $J = 2$ are $2_{0,2} \rightarrow 2_{1,1}$, $2_{1,1} \rightarrow 2_{2,0}$, and $2_{1,2} \rightarrow 2_{2,1}$. The assignment of the rotational spectrum of an asymmetric rotor is more complicated than that of a linear or symmetric-top molecule because of the greater number of possible transitions and the lack, in many cases, of any regular spectral pattern (see Fig. 2). Unlike linear and symmetric-top molecules, transitions between high J levels can fall at low frequencies for an asymmetric top.

The general features of the rotational energy levels and spectra discussed previously will be modified somewhat when the effects of nonrigidity, nuclear coupling, etc., are taken into account. One effect of nonrigidity, centrifugal distortion, has already been mentioned. A real molecule is vibrating as well as rotating even in the ground vibrational state; the rotational constants will be different for each vibrational state v, and for each state v a separate rotational spectrum is obtained. Because of intensity considerations, however, not all of these lines will be observable. The most important type of nuclear interaction arises when the molecule has one or more nuclei with a nonzero nuclear quadrupole moment. The interaction of the nuclear electric quadrupole moment with the molecular electric field gradient at the nucleus gives rise to a nuclear quadrupole hyperfine structure; i.e., each rotational transition is split into a number of components. Application of external electric or magnetic fields also perturbs the rotational energy levels and therefore affects the rotational spectrum. The more important is the effect of electric fields, commonly called the Stark effect. The electric field interacts with the molecular dipole moment, and as a result each rotational line will split into a number of components. To aid in the detection of rotational spectra, the Stark effect is usually used to modulate rotational lines. This forms the basis of the Stark-modulation spectrometers which will be discussed in Section 1.4.

Because it is only the location and intensity of a line belonging to a

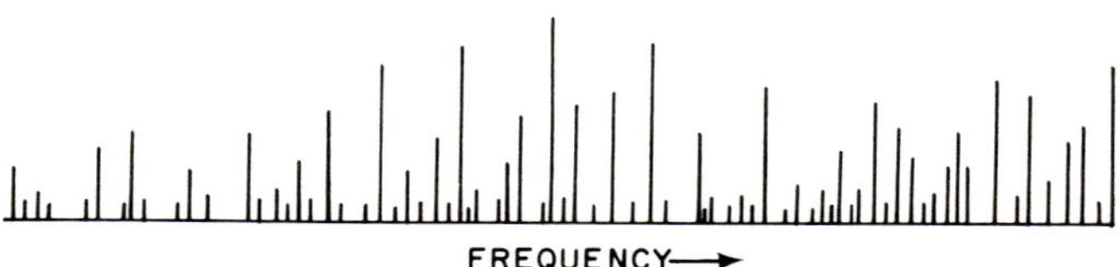

FIG. 2. General pattern of the rotational spectrum of an asymmetric rotor.

particular molecule which are important for chemical analysis, a detailed analysis of the spectrum is not required. Further discussions of interaction effects in rotational spectra may be found in the books cited previously.

1.2 Line Shape

As the radiation source is tuned to a particular frequency where the molecules will absorb [i.e., a frequency ν for which $\nu = (E_f - E_i)/h$], there will be a decrease in the power reaching the radiation detector. The power P reaching the detector depends on the absorption coefficient α_ν of the gas at the frequency ν and the attenuation constant of the cell α_c. Specifically,

$$P = P_{in} \exp[-(\alpha_\nu + \alpha_c)l] = P_0 \exp[-\alpha_\nu l] \tag{4}$$

where P_{in} is the input power to the absorption cell, l is the radiation path length, and P_0 is the microwave power at the detector with no gas in the cell. The power absorbed by the gas, $\Delta P = P_0 - P$, is given by

$$\Delta P = \alpha_\nu l P_0 \tag{5}$$

where it is assumed $\alpha_\nu l \ll 1$. This is valid under the usual experimental conditions.

Absorption lines are never perfectly sharp, but rather absorption takes place over a band of frequencies, giving rise to a line width. The line width $2(\Delta\nu)$ is defined as the width in frequency units between half-intensity points; the half-width is therefore designated by $\Delta\nu$. The line width sets an upper limit on the resolution obtainable. The various factors which contribute to the line width in rotational spectra will be considered below.

1.2.1 *Natural Line Width*

The natural line width arises from the probability of spontaneous emission, which leads to a finite spread of the upper level. This width is negligibly small ($<10^{-4}$ Hz) for the usual observations of rotational spectroscopy.

1.2.2 *Doppler Broadening*

At low pressures, the principal contributor to the line width is the Doppler effect. Since some molecules are moving with velocity components parallel or antiparallel to the radiation, the frequency of absorption will be shifted slightly. The absorption frequencies of those moving perpendicularly are unaffected. The velocity distribution will be a Maxwell–Boltzmann distribution, and hence various Doppler shifts will result. The line shape will be Gaussian in nature, and the line width is given by

$$2(\Delta\nu) = 7.15\, \nu_0 (T/M)^{1/2} \times 10^{-7} \tag{6}$$

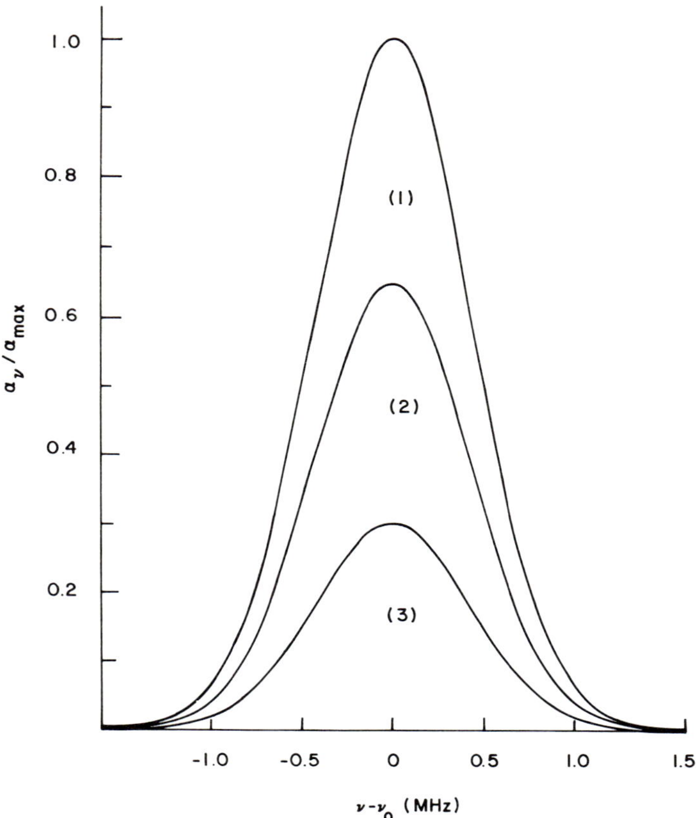

FIG. 3. Line shape of Doppler-broadened lines. $\Delta\nu = 0.50$ MHz. Values for p/p_c: curve (1), 1; curve (2), 0.65; curve (3), 0.30. [From W. Gordy and R. L. Cook (1970). "Microwave Molecular Spectra," Part II of Chemical Applications of Spectroscopy, 2nd ed., Wiley, New York. Used with permission.]

where ν_0 is the resonance frequency, T is the absolute temperature of the gas, and M is the molecular weight. The Doppler width is thus independent of pressure and increases with T and ν_0. In the centimeter wave region, line widths less than 100 kHz are typical, whereas in the submillimeter wave region, line widths of over a megahertz can be expected.

In Fig. 3 the line shape for Doppler-broadened lines is illustrated. In this pressure region, the peak intensity increases linearly with pressure, while the line width remains constant. As the pressure is raised, a point is reached where pressure broadening due to molecular collisions starts to become important. At this pressure, $p_c \simeq 1$ mTorr (1 mTorr $= 10^{-3}$ mm of Hg $=$ 1 μm), further increase in the gas pressure leads to an increase in the line width.

1.2.3 *Pressure Broadening*

In the pressure region $p \gtrsim 10$ mTorr, collisions between molecules become the important factor determining the line width. If τ is the mean time between molecular collisions, the half-width is given by

$$\Delta\nu = 1/2\pi\tau \tag{7}$$

At pressures not too high, the line shape is Lorentzian and is illustrated in Fig. 4. This condition holds over a reasonably large range of pressures, and it is in this region where most transitions are measured, since this gives the maximum peak absorption. Furthermore, this peak intensity is independent of pressure (see Section 1.3). In this region, since the collision time which broadens the line is inversely proportional to pressure, the line width increases linearly with pressure. Typically, $(\Delta\nu/p) = 10$ MHz/Torr;

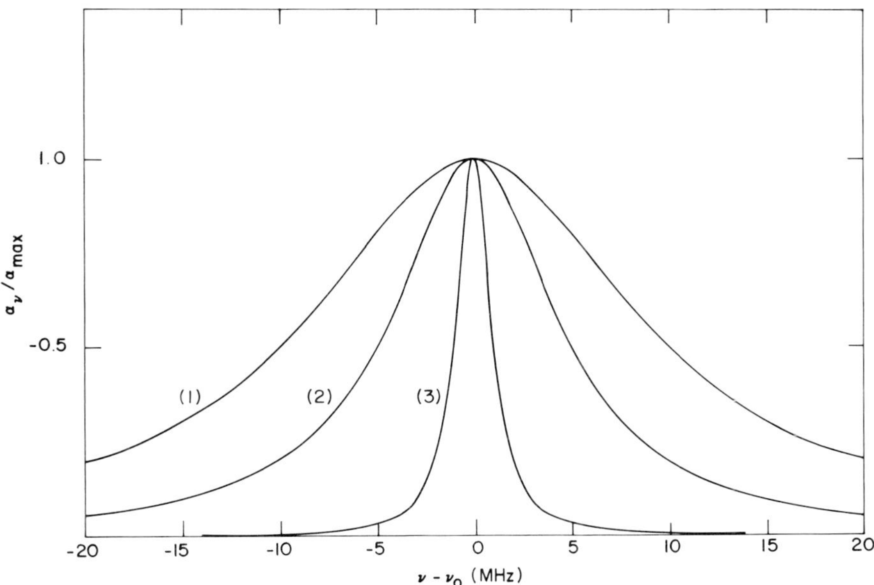

FIG. 4. Line shape of pressure-broadened lines. The half-width constant $(\Delta\nu)_1 = (\Delta\nu/p)$ is typical. $(\Delta\nu)_1 = 10$ MHz/Torr. Values of p: curve (1), 1 Torr; curve (2), 5×10^{-1} Torr; curve (3), 10^{-1} Torr. [From W. Gordy and R. L. Cook (1970). "Microwave Molecular Spectra," Part II of Chemical Applications of Spectroscopy, 2nd ed. Wiley, New York. Used with permission.]

therefore at $p = 10$ mTorr, $2\Delta\nu = 200$ kHz, and for $p = 100$ mTorr, $2\,\Delta\nu = 2$ MHz. Although the half-width constant $(\Delta\nu/p)$ is essentially a constant for a given gas within this pressure range, it may in fact be a slightly different constant for different rotational transitions. The approximate temperature dependence of $\Delta\nu$ is $1/T$. Molecules with large dipole moments have large $\Delta\nu$. For best resolution low pressures should be employed; however, this will sacrifice sensitivity, and some compromise must usually be chosen.

1.2.4 *Instrumental Broadening*

When the mean free path is comparable with any dimension of the cell, cell-wall broadening can be important. At sufficiently low pressure where collisions with the wall are dominant, the line width depends mainly on Doppler broadening and cell-wall broadening. The cell-wall broadening half-width, $\Delta\nu = 1/2\pi\tau$, depends on the mean time τ between collisions of the molecules with the cell walls. The line width can be expressed (in kilohertz units) by (Danos and Geschwind, 1953; Townes and Schawlow, 1955)

$$2(\Delta\nu) = 2.3[(1/a) + (1/b)](T/M)^{1/2} \tag{8}$$

where a and b are the cross-sectional dimensions (in centimeter units) of a long rectangular cell. For the usual X-band Stark cell, the dimensions ($a = 0.2$ in. $\ll b$) are such that $2(\Delta\nu)$ is smaller (<25 kHz) than Doppler broadening.

If the gas sample is subjected to too much microwave power, the thermal equilibrium distribution of the states will be upset, and saturation effects are caused which affect the line shape and absorption coefficient. Such power saturation becomes important at low pressures for power levels of only fractions of a milliwatt. It may be avoided by increasing the pressure or reducing the power. Often the effects of power saturation cannot be completely avoided, because if the power level is lowered too much, the detection efficiency of the crystal detector drops with a corresponding loss of sensitivity. The line width is increased (Townes and Schawlow, 1955) by power saturation and is given by

$$2(\Delta\nu) = 2\Delta\nu_0(1 + KP_0)^{1/2} \tag{9}$$

where $\Delta\nu_0$ is the half-width for the unsaturated line, and K is the power saturation coefficient [see Eq. (16)].

Modulation of spectral lines by the use of the Stark effect or by frequency modulation of the radiation source causes line broadening. This effect is reduced by keeping the modulation frequency less than the line width.

1.3 Absorption Coefficient and Line Intensity

In the collision broadening pressure range (10^{-2} Torr to approximately 1 Torr) where $\nu_0 \gg \Delta\nu$ and the line shape is Lorentzian, the absorption coefficient for the transition $J, \tau \rightarrow J', \tau'$ is expressed by

$$\alpha_\nu = \frac{8\pi^2 N F_{J,\tau} \nu^2 \mu_g{}^2 \lambda_g(J,\tau; J',\tau')}{3ckT(2J+1)} \left[\frac{\Delta\nu}{(\nu_0 - \nu)^2 + \Delta\nu^2}\right] \tag{10}$$

where N is the number of molecules per unit volume; $F_{J,\tau}$ the fraction of molecules in the lower state J,τ of the transition; ν the frequency of the microwave radiation; ν_0 the resonant frequency, frequency of maximum absorption; $\Delta\nu$ the half-width of the line [when $(\nu_0 - \nu) = \pm\Delta\nu$, α_ν is half of its maximum value]; μ_g the electric dipole moment component giving rise to the particular transition under observation; $\lambda_g(J,\tau;\ J',\tau')$ the line strength of the transition, which is related to the square of the transition moment; c the speed of light; k the Boltzmann constant; and T the absolute temperature of gas at thermal equilibrium. Here τ has its usual meaning for an asymmetric top; for a symmetric top $\tau = K$, and for a linear molecule τ is omitted.

The line strengths for asymmetric rotors have been tabulated for various values of κ (Wacker and Pratto, 1964). For linear and symmetric top molecules, where the dipole moment is associated with only one principal axis $\mu_g \equiv \mu$, simple expressions apply. In particular,

$$\lambda(J,K \rightarrow J+1,K) = [(J+1)^2 - K^2]/(J+1) \tag{11}$$

for symmetric-top molecules. This also applies to a linear molecule with $K = 0$. The quantity $F_{J,\tau}$ for all classes of molecules is given by

$$F_{J,\tau} = F_v[g(2J+1)\exp(-E_{J,\tau}/kT)/Q_r] \tag{12}$$

where $E_{J,\tau}$ is the rotational energy of the lower state J,τ. The quantity g is a statistical weight factor which accounts for the twofold degeneracy in symmetric-top $K \neq 0$ levels and for the effects of nuclear spin statistics on the rotational energy level populations. The quantity Q_r is the rotational partition function, and its definition depends on the type of molecule. It is a function of the temperature and the rotational constants. The quantity F_v is the fraction of molecules in the vibrational state v being observed. This may be evaluated if vibrational frequencies are known. In the ground state, at low temperatures $F_v = 1$ is often a reasonable approximation. Discussions of the evaluation of the various quantities in Eq. (12) have been given (Townes and Schawlow, 1955; Gordy and Cook, 1970). It may be noted that the actual evaluation of α_ν from molecular parameters will usually not be required for chemical analysis.

A particular useful quantity is the peak absorption coefficient, i.e., the absorption coefficient at the resonant frequency $\nu = \nu_0$. If we let p denote the total pressure in the absorption cell and x be the mole fraction of the particular molecular species being considered, then $N = xp/kT$, and we may write

$$\alpha_0 = [8\pi^2 F'_{J,\tau}\nu_0^2\mu_g^2\lambda_g(J,\tau;J',\tau')/3ck^2T^2(\Delta\nu/p)]x \tag{13}$$

where $F'_{J,\tau} = F_{J,\tau}/(2J+1)$. Since $\Delta\nu$ is proportional to p, the quantity $(\Delta\nu/p)$ is a constant. It therefore follows that α_0 is independent of the total pressure (see Fig. 4) and varies linearly with the mole fraction x. However, as discussed in Section 2.2.2, the dependence on the concentration is generally not so simple, but depends on the sample composition because $\Delta\nu$ does.

For quantitative analysis the integrated line intensity, i.e., the intensity integrated over the whole line, is useful and is defined by

$$\alpha_{\text{int}} = \int_0^\infty \alpha_\nu \, d\nu = \pi\alpha_0\Delta\nu \tag{14}$$

where α_ν is given by Eq. (10). Either the measurement of the area under the absorption curve or the measurement of α_0 and $\Delta\nu$ will give α_{int}. From the definition of α_0, it is apparent that α_{int} is independent of the line width $\Delta\nu$ and is dependent on the partial pressure of the absorbing species.

Measurement of α_0 or α_{int} requires operation at sufficiently low radiation power levels to avoid power saturation. This is often inconvenient because of the low signal levels incurred. The effect of power saturation is to decrease α_0 and is given by (Townes and Schawlow, 1955; Bleaney and Penrose, 1947)

$$\alpha = \alpha_0/(1 + KP_0) \tag{15}$$

where the power saturation coefficient is given by (Harrington, 1967)

$$K = (\lambda_g/\lambda)(64\pi^3 \,|\mu_{ij}|^2 t\tau/3h^2cA) \tag{16}$$

Here K is in cgs units; λ_g/λ is the ratio of the waveguide to free space wavelength, $|\mu_{ij}|$ the dipole transition matrix element, A the cross-sectional area of the cell, $\tau = 1/2\pi\Delta\nu$, the mean time between collisions which broaden the line, and t is the mean time between collisions restoring equilibrium. The saturation coefficient depends on the particular molecule, the transition studied, sample pressure, and composition, among other things.

Harrington (1967) has recently proposed a new intensity coefficient Γ which makes use of the phenomenon of power saturation and is useful in quantitative analysis. This coefficient is defined by

$$\Gamma = \Delta P/lP_0^{1/2} \tag{17}$$

which is similar to the definition of α given in Eq. (5), although it has a different dependence on P_0. Making use of Eqs. (17) and (15), the Γ absorption coefficient at the resonant frequency can be written in the form

$$\Gamma = \eta\phi \tag{18}$$

where

$$\eta = \alpha_0/K^{1/2} \tag{19}$$

From Eqs. (7), (13), and (16), and the assumption that each collision is effective in restoring an equilibrium distribution (that is, $t = \tau$), it is evident that η depends on the partial pressure xp of the absorbing species, but is independent of $\Delta\nu$. Hence η has the advantage of not depending on the composition of the sample as does α_0. The dimensionless power density function ϕ is always a function of the dimensionless product KP_0. Its explicit dependence on KP_0 depends on the power distribution in the cell:

$$\phi(KP_0) = (KP_0)^{1/2}/(1 + KP_0), \qquad \text{uniform power distribution} \tag{20}$$

$$\phi(KP_0) = (KP_0)^{-1/2}[1 - (1 + KP_0)^{-1/2}], \qquad \text{Stark cell } \mathrm{TE}_{10} \text{ mode power distribution} \tag{21}$$

The properties of these functions are similar. They increase with an increase in power rather rapidly to a broad maximum and finally decrease with further increase in power (see, e.g., Fig. 8).

Harrington (1967) has shown that for a Stark-modulated spectrometer Γ is proportional to the signal amplitude (signal voltage) provided the rectified crystal-detector current is kept constant. In this regard see also Crable and Wahr (1969). The maximum line signal therefore requires $\Gamma = \Gamma_{max}$, and this occurs at the power level for which $\phi = \phi_{max}$. Molecular concentration data contained in the new coefficient η can, therefore, be obtained under conditions of maximum signal, as

$$\Gamma_{max} = \eta\phi_{max} \tag{22}$$

1.4 Instrumentation

A fundamental microwave spectrometer consists of an energy source, an absorption cell, a detection system, and a system for measuring the source frequency. Reflex klystrons have seen the greatest use as energy sources. The output of klystrons is highly monochromatic. The output frequency can be changed by mechanical means or, for small frequency ranges, by applying a voltage to the reflector. In the past few years, a new type of microwave oscillator, the backward wave oscillator (BWO), has come into use. The BWOs have wide frequency ranges which can be scanned electronically. In addition, power levels can be controlled automatically. Both klystrons and BWOs can be automatically stabilized by electronic means.

The absorption cell usually consists of a rectangular waveguide with mica windows at each end to provide a vacuum seal. Small holes can be made in the waveguide for purposes of evacuation and sample introduction. Microwaves passing through the waveguide are detected by a crystal detector mounted in the end of the waveguide. The output voltage of the crystal is very small and must be amplified. The amplified signal is displayed on an oscilloscope or on a chart recorder. In the simplest spectrometers, absorption lines are displayed by sweeping the microwave source over a range of frequencies and observing the variations in power at the detector. Absorption lines show up as sharp dips on the oscilloscope or recorder tracing. For quantitative work a chart recorder is particularly useful.

A great increase in sensitivity can be achieved with the use of Stark modulation (McAfee *et al.*, 1949). For this purpose a thin metal septum is mounted in the cell parallel to the broad side of the waveguide using insulating material. An alternating (5–100 kHz) electric potential, usually in the form of a zero-based square wave, is applied to the septum. In the presence of an external electric field, the rotational energy levels and, therefore, the absorption lines are split. This results in a modulation (field on and off) of the absorption line at the frequency of the Stark modulator. The modulated signal is then detected and amplified by an amplifier tuned to the modulation frequency. The signal reaches the amplifier when the source being swept reaches the "field-off" absorption frequency or the frequency of the "field-on" absorption lines (Stark components). Phase-sensitive detection further increases the overall spectrometer sensitivity. Here the amplifier is referenced to the modulation frequency and only the noise that has the same frequency and phase as the signal is passed. By employment of phase-sensitive detection, the Stark components displayed in Fig. 5 may be distinguished (as illustrated in Fig. 15).

Frequencies of absorption lines in the microwave region are commonly measured with accuracies of one part in 10^6 or better. This is accomplished by comparing the frequency of the microwave source with the frequency of a stable oscillator. This oscillator, in turn, is calibrated with a frequency standard.

In 1965 Hewlett Packard introduced the first complete commercial microwave spectrometer, which is illustrated in Fig. 5. It uses a phase stabilized BWO source, Stark modulation (33.3 kHz), and a 2-m cell. The spectrometer has good frequency stability. Frequency measurements are made with an electronic counter. Frequency markers are displayed on a

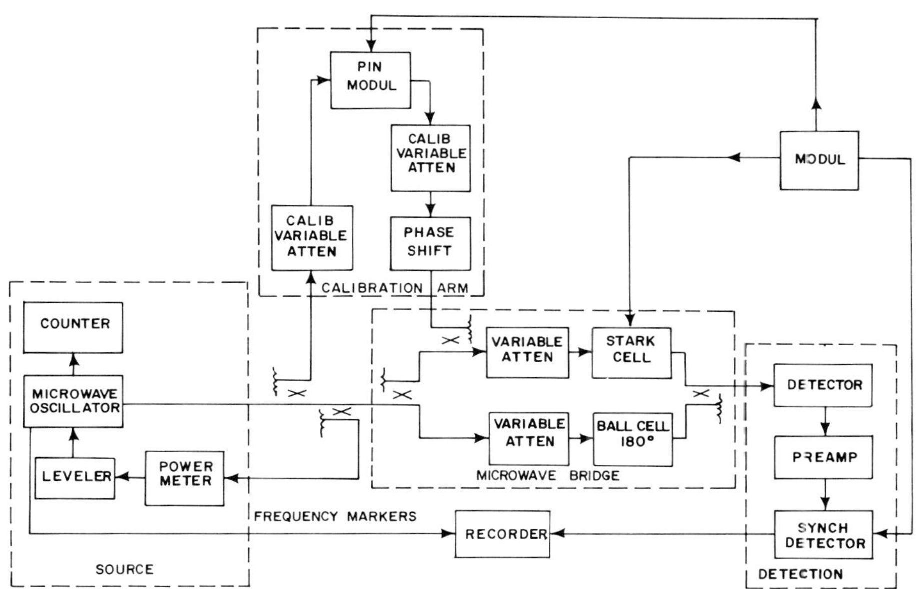

FIG. 5. Diagram of a microwave spectrometer employing a microwave bridge, phase-sensitive detection, Stark modulation, and automatic recording. Intensity measurements are made with the calibration arm. Frequency range is from 8.0 to 40.0 GHz. [Used with permission of Hewlett Packard, Palo Alto, California.]

chart recorder along with the absorption lines. Intensity measurements are made with a calibration arm fitted with precision calibrated attenuators. The calibration arm electronically generates a calibrated signal that simulates an absorption line. This calibrated signal can be compared to an absorption line, or it can be superimposed on the absorption line with a 180° phase shift to give a null measurement. The output dc voltage "signal," necessary for quantitative measurements, can also be obtained from the attenuator settings on the synchronous detector and the chart recorder displacement. Absolute intensity coefficient measurements can be made. A microwave bridge is used to maintain an optimum fixed rectified crystal current (~0.1 mA) while allowing the power in the absorption cell to be varied.

A Stark-modulated spectrometer with phase-sensitive detection, called Camspek, is also available from Cambridge Scientific Instrument Limited (Cuthbert *et al.*, 1971). The instrument covers the R-band region with a BWO, and employs a Stark-modulation frequency of 40 kHz and a 1-m sample cell. Designed specifically for analytical applications, Camspek incorporates a convenient sampling system. To facilitate sample removal, the sampling system and cell can be heated to 150°C. The sample cell is also easily dismantled for cleaning purposes. Quantitative analysis measurements are made by measuring the integrated absorption coefficient, α_{int}. Camspek uses an electronic digital integrator system to determine the line area.

For special problems where the concentration of a particular gas in a sample is to be monitored, such as in process control, a cavity spectrometer, which cannot be conveniently scanned over large frequency ranges, would be useful. Various cavity spectrometers have been described (Verdier, 1958; Beers, 1959; Dymanus, 1959; Lichtenstein *et al.*, 1963; Verdier and Wilson, 1958). Importantly, cavity spectrometers offer increased sensitivity by increasing the effective sample path length. Also, quantitative analysis using the Γ coefficient (see Section 2.2.3) requires power saturation of the absorption line, and this is more easily achieved in a cavity because of the higher power densities. A simple cavity spectrometer system could be constructed which, though less versatile, would offer the advantage of lower cost, sensitivity, and the portability required for remote monitoring. A cavity spectrometer designed to monitor trace gases has been discussed by Hrubesh (1971). The applicability of a cavity for the measurement of small amounts of gases in mixtures has also been discussed by Srinivasan (1973).

Further details on microwave instrumentation may be found in Gordy *et al.* (1953), Townes and Schawlow (1955), Sugden and Kenney (1965), and Wollrab (1967). A recent review of the techniques and instrumentation in the shorter millimeter and submillimeter wave region is given by Gordy (1965).

2 Applications and Limitations

2.1 Qualitative Uses

The particular characteristic molecular property which forms the basis of a microwave analysis is the rotational absorption spectrum. Since the rotational spectrum depends on the principal moments of inertia, the spectra obtained are characteristic of the whole molecule rather than certain parts of the molecule. This is in contradistinction to infrared absorption where the absorption frequencies are characteristic of certain functional groups present in the molecule. Small changes of structure or composition of the overall molecule, such as an isotopic modification of an atom, give rise to appreciable changes in the positions of the rotational lines. This extreme sensitivity of the spectrum to small structural characteristics, coupled with the high resolution and accurate frequency measuring techniques, which allow lines to be measured easily to an accuracy of 50 to 100 kHz, makes it highly improbable that different substances will have a number of identical line frequencies. The rotational line frequencies thus provide a positive method of identification of a substance. It is only necessary to have a catalog or microwave atlas of spectra of known compounds. Qualitative identification is then made by comparing the measured lines of the unknown substance with the frequencies in the microwave atlas.

Microwave chemical analysis is applicable to a host of problems. It is well suited to the analysis of pure compounds or the components of a mixture, and to the analysis of isotopic mixtures, chemical isomers, or rotational isomers. Since the instrumentation is electronic in nature, a microwave spectrometer can be readily adapted to be computer controlled. It would therefore be particularly convenient for monitoring and for such industrial applications as process control. The outstanding property of the method, namely, its unique identification ability, can find application where other methods are inappropriate. The major drawbacks to its application, besides the inherent sample limitations considered in Section 2.1.3, are the high instrument cost and sensitivity. Crable (1972), for example, reports that the sensitivity of a conventional Stark spectrometer to important air pollutants is below that of gas chromatography and mass spectrometry. This situation can be improved by use of a cavity spectrometer, as discussed by Hrubesh (1971). This consideration, however, must be taken into account if sensitivity is the principal concern. In general, the method is most appropriate for isotopic analysis, for analyses requiring high specificity, and the detection or concentration determination of molecules of specific interest. Its strength does not lie, generally, in the analysis of very general analytical problems involving complex mixtures.

Rather than being a collection of graphs, the microwave atlas is a table of numbers, namely, the transition frequencies of various pure substances. The digital nature of the atlas lends itself to computer applications and data handling techniques. Tables are presently available from the National Bureau of Standards containing the line frequencies for substances which have been reported in the literature (see Cord *et al.*, 1968). These tables are, however, not specifically designed for chemical analysis, and the tabulated lines are generally not the strongest. An atlas containing a number of strong lines of each molecule spread over a given frequency range would be particularly convenient. At present, such an atlas is not available, although a limited one has been given by Lide (1966). For a complex mixture where little is known of its possible composition, a rather complete catalog of the spectrum of each molecule would be convenient. Such a comprehensive catalog is still in the future, although rapidly developing computer-controlled spectrometers will probably be used for its compilation. It should not be inferred from the above comments that the application of microwave spectroscopy must await a more complete atlas. Many important molecules have well-known lines, and for relatively simple gas mixtures present tabulations will be satisfactory. In addition, applications of the considerations of the following section will aid in the study of many mixtures with present catalogs.

2.1.1 *Selectivity*

Identification of an unknown is made by finding which substance has tabulated frequencies corresponding to the observed frequencies of the unknown. Two considerations arise which bear on the selectivity of a microwave chemical analysis. First is the number of spectral positions available, and second is the number of lines that must be measured to obtain a reasonably certain identification of the substance.

If one considers the region from 10,000 to 40,000 MHz and assumes that lines 100 kHz apart may be resolved, it follows that there are 300,000 resolvable spectral positions. In fact, in just the R-band region between 26,000 and 40,000 MHz (one spectral region effectively covered by a single backward wave oscillator), there are 140,000 spectral spaces available, more than enough room for a large number of lines without serious overlap problems. For an analysis, of course, the molecules must absorb in the above spectral region. Most substances exhibit lines throughout the microwave region. Except for a very few exceptions (about 5% of all polar molecules), the R-band region would, for example, suffice. The spectral region available, in reality, is much larger than the limited range selected above and extends from roughly 7000 to 800,000 MHz.

In many analytical problems, a single frequency measurement would suffice to identify the substance. However, even though the inherent resolution is very high, the possibility of overlapping lines and the measurement accuracy ultimately dictate the number of lines which are required to make an unambiguous identification. The possibility of overlap becomes important as the spectrum from a given molecule becomes dense and as the number of substances that have to be considered as possible components of the gas sample to be analyzed becomes great. A study of the overlap problem has been made by Jones and Beers (1971). Figure 6 shows the average percentages of cases that a line with a given measurement accuracy would be overlapped, based on the assumption that the line belongs to any one of a set of 33 molecules which have a total of 10,000 measured absorption lines. If, for example, a single line of an unknown is measured to a precision of ± 0.01 MHz (with care such precision can often be obtained),

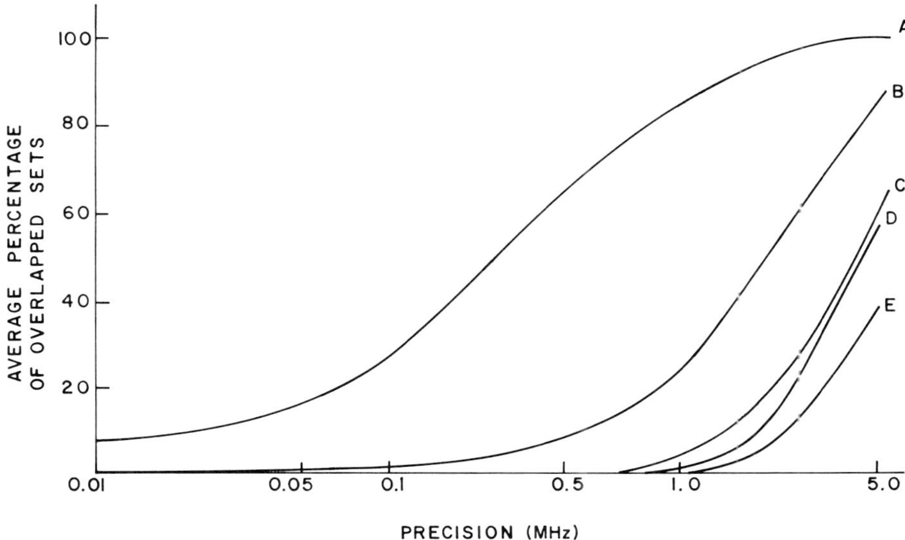

FIG. 6. Average percentage of overlapped sets of lines versus precision of measurement for a sample of 33 gases, plotted on a semilogarithmic scale. Each point on the curves represents an average of the percentage of overlap for ten molecules. Curve A: set of one line; B: set of two lines; C: set of three lines; D: set of four lines; E: set of five lines. [From G. E. Jones and E. T. Beers (1971). *Anal. Chem.* **43**, 656. Copyright (1971) by the American Chemical Society. Reprinted by permission.]

and if a microwave atlas indicates that a substance has a line cataloged at this frequency, one can be 96% confident that the unknown is that substance (curve A). On the other hand, if the line is weak, a measurement precision of 0.01 MHz may not be possible. A measurement precision of ±0.2 MHz would yield a confidence factor of only 55%. If, however, two lines were measured with the low precision of ±0.2 MHz (curve B), only a 2% chance of overlap is obtained, and hence, the confidence factor is now increased to 98%. The number of overlaps for a given measurement precision can be reduced if fewer possibilities have to be considered regarding the unknown. Assuming that the unknown must be one of only ten possible molecules, for example, the results of Fig. 7 are obtained. The drop in the percentage of overlaps is apparent. Usually, sufficient knowledge about the unknown sample will be available to limit the possibilities which need to be considered.

The results of Figs. 6 and 7 can be used as a rough guide in selecting the number of lines to be measured for the given degree of confidence desired.

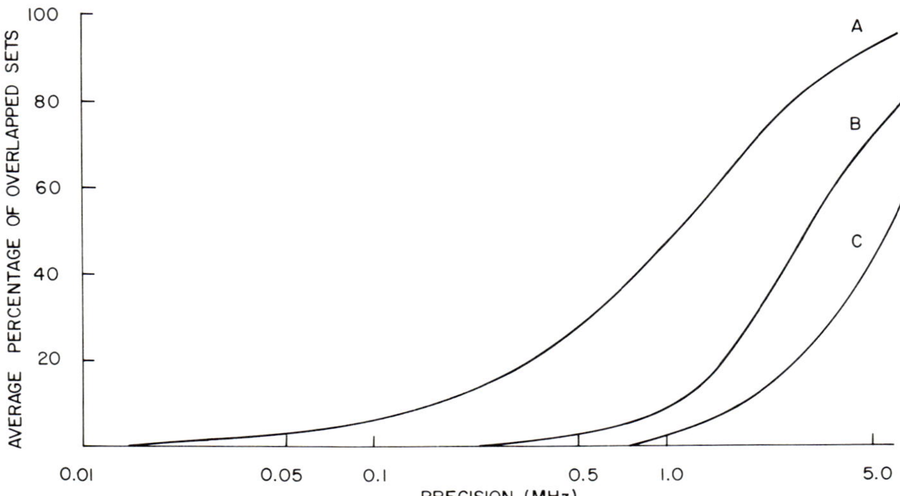

FIG. 7. Average percentage of overlapped sets of lines versus precision of measurement for a sample of 10 gases. Curve A: set of one line; B: set of two lines; C: set of three lines. [From G. E. Jones and E. T. Beers (1971). *Anal. Chem.* **43**, 656. Copyright (1971) by the American Chemical Society. Reprinted by permission.]

They suggest that an alternative to the measurement of a few lines (two or three lines) at high precision would be the measurement of more lines with less precision, which in the long run would be less time-consuming. In any case, we are talking about the measurement of a very small number of absorption lines.

Besides the employment of line frequencies, which would usually be obtained under conditions of high resolution, one can also use low-resolution spectra to distinguish and identify various substances (see Section 2.3). These low-resolution spectra are obtained with fast scan rates and may cover a whole microwave band or only a portion of it. With a scan rate of 10 MHz/sec, the whole R-band region may be covered in approximately 20 min. The appearance of such spectra is dependent on the Stark voltage, scan rate, and the pressure employed. Low-resolution spectra are particularly useful for molecules which give rise to some distinguishing spectral pattern. This can be the case, for example, for near prolate asymmetric tops ($\kappa \simeq -1$), with "a"-type R-branch transitions, or with "c"- or "b"-type Q-branch transitions, and for near oblate asymmetric tops ($\kappa \simeq 1$) with "c"-type R-branch transitions, or with "a"- or "b"-type Q-branch transitions.

2.1.2 *Sensitivity*

In the application of microwave spectroscopy to either qualitative or quantitative analysis, the major concern will be the ability to detect the spectrum of the molecule in question, assuming it meets the nominal requirements of vapor pressure and dipole moment. The ease of detectability depends on the absorption coefficient. The larger the absorption coefficient, the more readily one can observe small concentrations of the molecule (see also Section 2.1.3). Absorption coefficients vary widely among molecules and among different transitions of the same molecule. Typically, the weakest line that a good Stark-modulated microwave spectrometer can detect corresponds to an absorption coefficient α_0 of about 10^{-9} cm^{-1}. If one considers a cell of 2-m length and $\alpha_0 = 10^{-9}$ cm^{-1}, the fractional power absorbed would be, from Eq. (5), $\Delta P/P_0 = 0.0000002$. Hence, only 0.00002% of the power is absorbed at the peak frequency, and it is only the high sensitivity of microwave spectroscopy which allows the detection of such weak absorption lines. In the centimeter wave region, absorption coefficients are usually in the range of 10^{-5} to 10^{-7} cm^{-1} or smaller, although strong absorbers with coefficients greater than 10^{-5} cm^{-1} have been observed. In the shorter millimeter and submillimeter wave region, α_0's approaching 1 cm^{-1} are encountered.

Various factors affect the absorption coefficient and therefore the detectability. From the expression for α_0, it is apparent that by going to higher frequencies, α_0 will be increased. The shorter millimeter wave region is therefore very attractive because of the significant gain in the absorption coefficient. The choice of frequency range will usually be dictated by one's experience or the availability of equipment or both. At present, the only commercial microwave spectrometers sold as a complete system operate in the region from 8 to 40 GHz. This frequency region is also the region where most heavy molecules have been observed.

The dipole moment is of prime importance, since α_0 varies as the square of μ. Molecules with small dipole moments (<0.2 D) can be expected to have weak spectra. The line width, however, also decreases with the dipole moment. This effect helps for molecules with small dipole moments by increasing the peak absorption coefficient.

If possible, the pressure chosen should be such that one operates in the collision-broadened region where α has its maximum value. The usual operating range will be between 10 and 100 mTorr.

Usually, low temperature, e.g., that of dry ice, will be chosen if the sample has sufficient vapor pressure. This follows from the inverse temperature dependence of α_0 ($\alpha_0 \sim 1/T^2$ for a linear molecule, $\alpha_0 \sim 1/T^{5/2}$ for symmetric and asymmetric tops). Lowering of the temperature also helps to improve the F_v factor, since this decreases the population in the excited vibrational states leading to an increase in the ground-state line intensities. Significant improvement, however, will not be realized if a molecule has a number of very low-frequency vibrational modes. Similarly, the presence of hyperfine structure or the presence of naturally occurring isotopic modifications give rise to a decrease in the line intensity. The latter effect is important for nuclei such as Br where the abundances are $^{79}Br = 50.5\%$ and $^{81}Br = 49.5\%$. Such effects are particularly important if the molecule under study has a very small dipole moment. The presence of one or more of the above mentioned effects could reduce the intensity to an unobservable level.

The absorption coefficient also depends on the size of the molecule through the partition function Q_r. Large molecules with large moments of inertia, and therefore large partition functions, will have weaker absorptions, other things being equal.

Very large molecules usually possess most of the disadvantages mentioned above so that microwave spectroscopy will be of limited use for them.

The line strength factor is a function of the rotational quantum numbers; for asymmetric tops, it also depends on κ. It increases with J for linear and symmetric-top molecules and for certain asymmetric top transitions, although the general behavior with J is more complicated for the latter. As a general guide, it may be stated that the moderately high J transitions will be stronger than the low J transitions. At sufficiently high J, the exponential factor in Eq. (12) will override the gain from λ. The λ factor has the same value for the same transition in different linear or symmetric top molecules. For asymmetric tops, this would be true only if κ is also the same.

2.1.3 *Size and Kind of Specimen*

To be suitable for study by microwave spectroscopy, the substance must have a nonzero electric (or magnetic) dipole moment and sufficient vapor pressure at the temperature of observation. A minimum vapor pressure of approximately 1 mTorr is required. Many liquid substances and even solids have sufficient vapor pressure for observation at room temperature. In addition, by use of specially designed high temperature cells, many substances can be heated to obtain the vapor pressure required. Ordinary Stark cells can usually be heated to 100°C without damaging the cell.

The amount of sample required is very small. A typical 2-m Stark cell (area $= 0.4 \times 0.9 = 0.36$ in.^{2}; volume $\simeq 470$ ml) filled to a moderate pressure of 50 mTorr at $T = 300°K$ requires only about 1.5×10^{-6} moles of gas. For a molecular weight of 100, this corresponds to a sample size of only 0.0002 g. The minimum concentration of a substance which can be detected in a gas mixture depends on the absorption coefficient. For a strong line, $\alpha_0 = 10^{-3}$ cm^{-1}, and taking 10^{-9} cm^{-1} as an upper limit of detectability, a mole fraction of as little as 10^{-6} could be observed. Thus if only 1.5×10^{-12} moles of a strong absorber were present in a gas mixture having a total pressure of 50 mTorr, the line could be detected. Hence, in favorable cases a small concentration of a substance, 1 ppm, can be analyzed in a sample mixture. Such small sample requirements enable microwave spectroscopy to be applied to microchemical analysis problems. For a typical weak line with $\alpha_0 = 10^{-7}$ cm^{-1}, a concentration of about 10,000 ppm (1%) would be required for observation. If sufficient information is available or can be estimated, a calculation of α_0 will indicate the minimum concentration required.

In gas samples which may contain substantial quantities of nonpolar gases, e.g., CO_2, N_2, O_2, these should be separated from the sample before a microwave analysis is carried out. Such techniques of sample enrichment are particularly important for weak absorbers in order to obtain sufficient concentration for detection.

2.1.4 *Nondestructive Test*

Since the radiation in the microwave region is nondestructive, the sample is unaffected and may be recovered in its original structural and chemical state. However, with a Stark-modulated spectrometer, metal cells are usually employed with some insulation on the inside for support of the Stark septum, so that some problems with adsorption and chemical reactivity in the cell can arise for certain molecules. Chemical reactivity can present serious problems if the lifetime in the sample cell is not sufficient for observation of the spectrum. Sometimes the reaction rate can be reduced by "conditioning" the cell. This involves filling the cell with successive samples at somewhat high pressure. For very reactive species, a flow system may be employed. Fresh sample is bled in at one end of the cell and pumped continuously out at the other end of the cell. The flow rate is adjusted to obtain the desired pressure and renewal rate. The reaction can also, in many cases, be reduced sufficiently by cooling the cell. Special absorption cell design and construction materials can be employed for unusual problems (see, for example, Hardy *et al.*, 1954; Kewley *et al.*, 1963).

2.1.5 *Time Required for Analysis*

The time required depends on a number of factors, and it is difficult to specify an exact or even approximate analysis time. If some idea of the composition is available, then checks can be made at various frequencies where absorption is expected. If the lines are reasonably strong, such a procedure can be completed very rapidly. Three or four frequencies can be checked in approximately 10 min, depending on the spectrometer details. Alternatively, a portion of the accessible frequency range may be scanned rapidly and the stronger lines checked for assignment. One can then search for the components with weak lines or small concentrations. For weak lines, longer detector time constants and much slower scan rates are required to obtain maximum sensitivity, e.g., 0.02 MHz/sec. For a moderately slow scan rate of 0.1 MHz/sec, ~3 hr would be required to search over a frequency spread of 1000 MHz. The fast scan rates may be carried out at relatively high pressures. The resolution, however, is then poorer because of the larger line widths, and this will be a disadvantage if the observed spectrum is very dense. Also, at fast scan rates, the accuracy of the frequency measurements is reduced. Usually some compromise between pressure and scan rate must be selected to obtain the desired frequency measurement accuracy and resolution, and this is dependent on the characteristics of the observed spectrum. If the sample composition is completely unknown and weak lines are observed, it is obvious that much more time will have to be spent in the analysis.

2.2 Quantitative Uses

2.2.1 *Sensitivity*

Quantitative analysis by microwave spectrometry can be performed only on constituents that have been identified. The absorption line to be used to determine abundance must have been identified as belonging to a certain species. That is, quantitative analysis by microwaves presupposes a qualitative analysis. Consequently, some of the considerations to be made in a determination of concentrations are identical to those already discussed in connection with qualitative analysis (see Section 2.1). The previous considerations of sample size and selectivity are unchanged. As no additional operations are performed on the sample, the quantitative analysis is nondestructive, just as were qualitative determinations.

In principle, an absorption line with a signal-to-noise ratio of just over 1 to 1 could be used to identify a molecule. Lines to be used to determine concentrations must have signal-to-noise ratios higher than 1 to 1 to give concentrations with fair accuracy. In practical cases, however, molecules will rarely if ever be identified exclusively by lines with signal-to-noise ratios near 1. It is very likely that the strongest transition that was used to identify the molecule is sufficiently intense to allow a concentration determination. Therefore, the previous discussion of sensitivity applies in practical cases to quantitative as well as qualitative analyses.

2.2.2 *Use of the α Coefficient*

From Eq. (5) the following expression for the Beer's law intensity coefficient α may be written:

$$\alpha = \Delta P / l P_0 \tag{23}$$

The radiation path length l is related to the length of the absorption cell by $l = (\lambda_g/\lambda)L$, where λ is the wavelength in free space corresponding to the microwave frequency, and λ_g is the wavelength of the microwaves in the waveguide. The coefficient α_0 is obtained experimentally by measuring the quantities on the right side of Eq. (23) when the spectrometer is tuned to the resonant frequency of an absorption line.

The concentration of the gas causing the absorption can, in principle, be determined with measurements made on only one absorption line. From Eq. (13),

$$\alpha_0 \Delta\nu = C \nu_0^2 x p \tag{24}$$

and a determination of the partial pressure xp requires, in addition to measurements of α_0 and $\Delta\nu$, a knowledge of the molecular constants in order to calculate the constant C. A measurement of the total pressure p then gives the mole fraction x. This procedure will be useful only in a small minority of cases, however.

There are other methods of determining abundances using absolute measurements of α_0 that do not require accurate molecular information. Owing to difficulties in obtaining accurate measurements of α_0, $\Delta\nu$, and p, these methods have come into use only with the advent of reliable commercial spectrometers and reliable pressure gauges of the appropriate range. Before describing intensity measurements with these spectrometers, let us first discuss methods used prior to their introduction. These methods make use of relative intensity measurements.

An obvious approach to the problem of determining concentrations, which does not require an absolute measurement of the intensity coefficient, relies on comparisons of line intensities in the unknown mixture to those in a mixture whose composition is known. Fortunately, measurement of signals or peak heights is both simple and accurate. The ratio of the output signals obtained from a Stark spectrometer is equal to the ratio of the absorption coefficients provided the microwave power levels at the two absorption frequencies and the effective path lengths are equal (Baird and Bird, 1954). Below a few hundred microamperes of crystal current, the power is proportional to the rectified crystal current, so that the power level can be held constant by monitoring the detector crystal current. According to Eq. (24), α_0 is proportional to the concentration x for a given absorption line provided the factor $\Delta\nu/p$ can be considered constant. Then with lines at the same frequency, the ratio of the peak heights is equal to the ratio of the concentrations. Abundances can then be determined by the simple procedure of comparing the signal of an absorption line of a molecule in the unknown mixture with the signal produced at the same frequency by the same molecule in a sample of known composition.

In general, however, $\Delta\nu/p$ is not constant but depends on sample composition, and serious errors can result if this factor is ignored. The mole fraction is actually proportional to $\alpha_0\ \Delta\nu/p$ rather than to α_0. The line width depends on the effective cross section of the molecule, which in turn depends on intermolecular interactions. For example, at a given total pressure, $\Delta\nu$ for a transition due to a molecular species diluted with a mixture of polar gases is likely to be much larger than $\Delta\nu$ for the same transition if the gas is diluted with nonpolar gases.

There are several procedures available for correcting this source of error. If one has sufficient knowledge of the unknown sample, a standard sample of known concentration with approximately the same composition as the unknown can be prepared. This is the simplest technique and is especially practical when a large number of determinations is to be made. Furthermore, it offers accuracies of a few percent if the signals are reasonably strong. Another approach is to dilute the unknown sample with a nonpolar gas such as nitrogen or helium for which line-width information has been previously obtained [see Eq. (26)]. This is feasible only if the line is relatively strong (see Section 2.1.3). The line-width information on the diluent is unnecessary if the standard sample is also diluted with the same diluent, as $\Delta\nu/p$ will then be very nearly the same for both samples. Finally, one can measure $\Delta\nu/p$ for the same transition in both the standard and the sample and use the exact equation

$$\frac{x_1}{x_2} = \frac{(\alpha_0)_1}{(\alpha_0)_2}\,\frac{(\Delta\nu/p)_1}{(\Delta\nu/p)_2} = \frac{(\alpha_{\mathrm{int}}/p)_1}{(\alpha_{\mathrm{int}}/p)_2} \tag{25}$$

where $(\alpha_0)_1/(\alpha_0)_2$ is given by the ratio of the peak heights.

It is because of the difference in the collision diameters of different types of molecules that α_0 in a mixture is not simply proportional to the mole fraction of the absorbing species. Assuming the broadening effects of the various constituents are additive, we can write (Hughes, 1952; Gordy, 1956)

$$(\Delta\nu)_{\mathrm{A}} = (k_{\mathrm{AA}}x_{\mathrm{A}} + k_{\mathrm{AB}}x_{\mathrm{B}} + k_{\mathrm{AC}}x_{\mathrm{C}} + \cdots)p \tag{26}$$

where $(\Delta\nu)_{\mathrm{A}}$ is the half-width for the measured transition of constituent A; x_{A}, x_{B}, x_{C}, etc., are the mole fractions of the various constituents; p is the total pressure; and k_{AA}, k_{AB}, k_{AC}, etc., are the collision broadening factors of A due to constituents A, B, C, . . . , respectively. A similar equation exists for the line widths of each constituent. There are as many equations of the above type as there are mole fractions, so that if the k's were known, measurements of line widths would make it possible to calculate all the concentrations. Alternatively, from Eq. (24), one could use the equations

$$(\alpha_0)_A = (C_A x_A)/(k_{AA}x_A + k_{AB}x_B + k_{AC}x_C + \cdots) \qquad (27)$$

Unfortunately, not many collision diameters have been accurately measured so that this approach offers limited usefulness at present. It may be noted that if a diluent B is the major constituent, only information on k_{AB} is required. Furthermore, diluting gas A with a second gas B reduces α_0 and increases $\Delta\nu$. For the isotopic species of a molecule $k_{AA} = k_{AB} = k_{AC} = \cdots$. Hence, $\Delta\nu/p$ is independent of the isotopic composition and is the same for each isotope. Also, C in Eq. (24) is expected to be essentially the same for the various isotopic species. Thus if variations in l are negligible and if the line frequencies are not too different, the isotopic concentration ratio x_1/x_2 will also be given by the ratio of the peak signal heights.

Except for a few instruments designed especially for the determination of α_0 (Bird, 1954; Mattuck and Strandberg, 1953; Dymanus *et al.*, 1960), most research spectrometers are unsuited for accurate absolute measurements of intensity coefficients. The principal problem in determining α_0 is the measurement of a very small microwave power change produced by absorbing molecules. For very low power levels, power is proportional to rectified crystal current, so that α_0 can be obtained from Eq. (23) if small changes in crystal current or power are measured. Therefore, accurately calibrated detector characteristics relating the output voltage to crystal current are required for absolute α_0 measurements by conventional spectrometers. In addition to the difficulty in measuring α_0, accurate measurements of $\Delta\nu$ can be difficult, particularly for very narrow lines. The precision with which frequency markers can be placed at the correct points on the absorption line limits the accuracy of $\Delta\nu$ to a few percent. A discussion of line-width measurements and a spectrometer designed for such measurements is given by Rinehart *et al.* (1965). Furthermore, pressure measurements in the region from 1 to 50 mTorr have been difficult until recently. The MKS Baratron capacitance manometer now seems to perform satisfactorily in this region (MKS Instruments, Inc., Burlington, Mass.).

Unlike the peak absorption α_0, the integrated intensity α_{int} does not depend upon the line width, but is directly proportional to the partial pressure; see Eq. (14). The most straightforward way to measure the integrated intensity coefficient is to measure the area under the absorption curve. In principle, one must integrate from zero to infinite frequency, but in practice one would measure on either side of the maximum only to the points at which the curve closely approaches the base line. Alternatively, one could measure the area from the maximum out to the points marking, for example, the half-width and multiply by the factor 4 (Hughes, 1952). Insufficient use of this technique has been made to allow reliable estimates of the accuracy of this approach; however, Crable (1972) has used area measurements to obtain concentrations to within a few percent.

Commercial spectrometers now make possible convenient and accurate measurements of absolute values of α_0 and α_{int}. Measurement procedures for α_0 are given in the Hewlett Packard Operating Manual. This has also been discussed by Curl (1969), who suggested relocating the calibration arm to simplify the measurement of α_0. Because of the high stability and reproducible characteristics, these spectrometers should also provide good line-width measurements. Concentrations can be determined by first measuring α_0, $\Delta\nu$, and p for a given transition in an unknown sample. The corresponding quantities are then measured for the same transition in a sample of known concentration. The unknown concentration is calculated from Eq. (25).

The Camspek spectrometer provides for the convenient measurement of α_{int} by means of an automatic electronic integration system, with provisions for detecting and correcting for line shape distortions from unresolved Stark components, etc. The concentration may be evaluated from Eq. (25.)

It must be remembered that the validity of Eq. (24) rests on the assumption that the pressure in the absorption cell is sufficiently high that pressure broadening dominates other line-broadening mechanisms. A second consideration that must be made with measurements based on the Beer's law coefficient α_0 is that the microwave power level must be sufficiently low that saturation effects are negligible. Otherwise the relationship between α_0 and concentration which was discussed previously is no longer valid. This second consideration can cause a serious inconvenience when the absorption line is relatively weak. From Eqs. (17) and (23), $\Gamma = \alpha_0 P_0^{1/2}$ in the unsaturated region (where $KP_0 \ll 1$). Since Γ is proportional to the spectrometer signal, we see that low power levels result in low signal levels and consequent loss of accuracy. This problem is circumvented by a completely different approach to quantitative analyses. This approach, based on Harrington's intensity coefficient Γ, actually takes advantage of the phenomenon of power saturation.

2.2.3 *Use of the* Γ *Coefficient*

The coefficient Γ is defined in Eq. (18) as the product of η and ϕ (KP_0). The advantage of the Γ coefficient is that it effects the separation of the line-width factor $(\sim 1/\tau)$ and the molecular concentration and, furthermore, allows one to work with the optimum spectrometer signal. The quantity K depends on τ and therefore on sample composition and total pressure. The power density function ϕ, however, is always a function of the product of K and P_0, so that any particular point on the ϕ curve corresponding to a fixed value of KP_0 can be reached by adjusting P_0 appropriately. An easily identifiable point on the curve of ϕ versus P_0 is the maximum value of ϕ. The coefficient η is independent of P_0, so that maximizing ϕ by adjusting P_0 also gives the maximum value of Γ. Since Γ is proportional to the signal, this choice of ϕ corresponds to the maximum signal amplitude. The general behavior of Γ versus power is shown in Fig. 8. According to Eq. (19), $\eta \sim (\tau/t)^{1/2}xp$ for a given transition. Hence, as long as the ratio (τ/t) is constant, η or Γ_{max} is proportional to the partial pressure. Preliminary evidence indicates that this is indeed the case, and usually $t \simeq \tau$ is assumed. This shows a difference between η and α_0, as α_0 is pressure independent. In summary, by adjusting the value of the

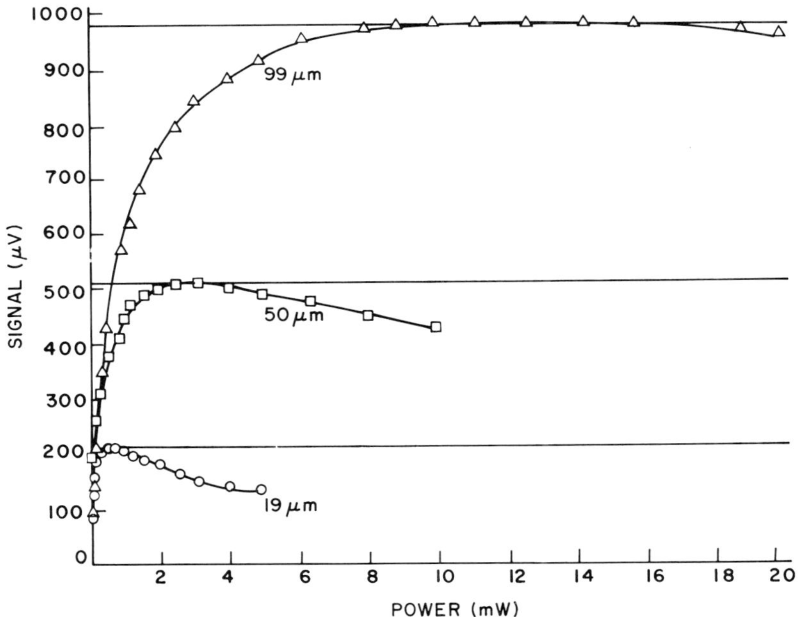

FIG. 8. Spectrometer signal for the $1_{0,1}-1_{1,0}$ transition in ethylene oxide as a function of the radiation power P_0. In each case the signal ($\sim\Gamma$) increases with increased power, finally reaching a broad maximum. The power required to reach the maximum signal is different for different pressures of pure sample. [From H. W. Harrington (1967). *J. Chem. Phys.* **46**, 3698. Used with permission.]

power such that Γ assumes its maximum value, the signal will be a maximum and proportional to the partial pressure of the particular species whose transition is being observed:

$$\Gamma_{\max} \sim S \sim xp \tag{28}$$

The coefficient Γ therefore allows a measure of the concentration which is independent of the line width. The linearity of the maximum signal amplitude with pressure for a pure sample ($x = 1$) of ethylene oxide is illustrated in Fig. 9. The effect of introducing an impurity gas while keeping the partial pressure of ethylene oxide at some fixed value is also shown. Note that the maximum signal amplitude does not change as the impurity gas is added.

A procedure for obtaining concentration using this coefficient would be to first use a sample of the pure gas to plot S versus pressure. The maximum signal S corresponding to the same transition of the gas in a mixture of unknown concentration is measured, and the calibration curve used to

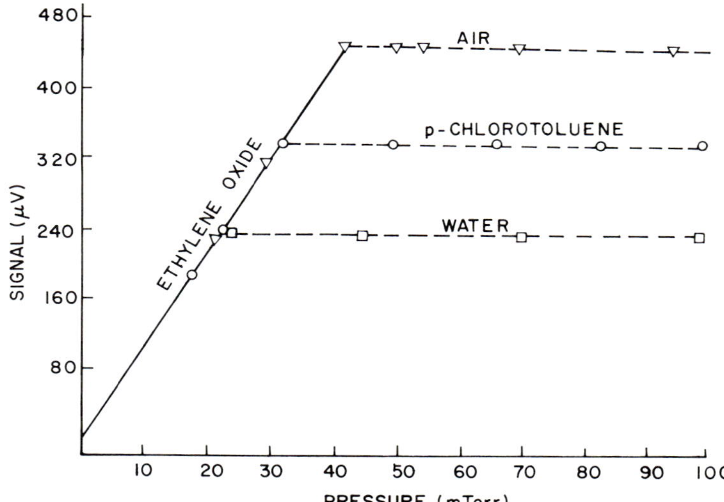

FIG. 9. The linearity of $\Gamma_{\max}$ with concentration of the absorbing species, ethylene oxide, and its independence of τ. The figure illustrates the results of three different runs using a different amount of ethylene oxide and a different impurity gas each time. The maximum signal is seen to increase linearly with increased pressure of the pure sample and does not change as an impurity gas is added while keeping the partial pressure of ethylene oxide constant. [From H. W. Harrington (1967). *J. Chem. Phys.* **46**, 3698. Used with permission.]

determine the partial pressure. A measurement of the sample pressure then gives the mole fraction x.

Actually, a calibration plot is not in principle necessary. According to Eq. (28), a plot of S versus the partial pressure should produce a straight line through the origin. One can therefore write for the mole fraction

$$x = (S_{\mathrm{m}}/S_{\mathrm{p}})(p_{\mathrm{p}}/p_{\mathrm{m}}) \tag{29}$$

where S_{m} is the maximum signal for the mixture, S_{p} is the maximum signal due to the same transition in the pure sample, and p_{p} and p_{m} are the total pressures of the pure gas and the unknown mixture, respectively. If a spectrometer with a calibration arm is used (see Section 1.4), the maximum signal S can be accurately obtained by using a simulated signal and a null method. The signal is then simply a reading in decibels from the precision attenuators. With S_{m} and S_{p} in decibels, then,

$$x = (p_{\mathrm{p}}/p_{\mathrm{m}}) \operatorname{antilog}[(S_{\mathrm{m}} - S_{\mathrm{p}})/20] \tag{30}$$

The above procedure, which promises a convenient and accurate method of determining abundances of gases in mixtures, depends upon the validity Eq. (28). Unfortunately, there has been experimental work (Rinehart, 1969) which indicates that a plot of maximum signal versus partial pressure does not pass through the origin as predicted. Specifically, it was found that the curve is a straight line through the origin for pressures up to 1 mTorr and is a straight line with a steeper slope for pressures above 3.5 mTorr. Between 1 and 3.5 mTorr the plot has a definite curvature. This departure from theory has not been entirely explained, but the discrepancy is believed due to the fact that collision broadening is not the only significant contribution to the line width. These considerations lead one to expect that the use of Eq. (30) would result in errors in measured concentrations. The magnitudes of these errors at various pressures, however, have not been determined. In any event, the less convenient approach of actually plotting the calibration curve should give correct answers even if the curve does not go through the origin. The pressure region below 3 or 4 mTorr should be avoided, however, as the assumption of the dominance of pressure broadening, as expected, is not valid here.

It should be remembered that the coefficient Γ is proportional to spectrometer signal if the rectified crystal current remains constant. For this reason, a spectrometer of the bridge-type is especially convenient (see Section 1.4), since the bridge allows the spectrometer to be operated at the necessary sample power levels with any desired crystal current. With a conventional, or "in-line," type of spectrometer, however, the crystal current is dictated by the power level at which $\Gamma_{\max}$ occurs. Nevertheless, using an in-line spectrometer with a calibration arm, Rinehart (1969) was able to overcome this difficulty by comparing the absorption signal of the gas with a reference signal produced by the calibration arm. This relative signal was found to be free of the effects of changes of crystal current and also provided a stable calibration curve. This curve, since it does not depend strongly on the idiosyncracies of a particular spectrometer, could perhaps be transferred between similar spectrometers. This calibration curve is very nearly a straight line in the pressure region above 3.5 mTorr. This curve still does not pass through the origin when extrapolated, however, so that reliable determinations of abundances cannot generally be made without a calibration curve as suggested by the theory and as previously described.

Preliminary investigations by Rinehart (1969) indicate that the most accurate results are obtained by still another procedure. A relative intensity is again defined as the ratio of the maximum intensity of an observed line to that of a standard reference line produced by the calibration arm. This relative signal is then plotted against the total pressure for the unknown mixture. Since S_{rel} is proportional to xp, this is equivalent to plotting Kxp versus p, where K is the constant of proportionality relating xp to S_{rel}. The plot of Kxp versus p is a straight line, the slope of which can be obtained from any two points to be

$$(Kxp_2 - Kxp_1)/(p_2 - p_1) = Kx \tag{31}$$

A similar slope can be obtained by plotting the relative signal versus total pressure for a standard sample in which the concentration of the absorber is a known value, x_s. The slope for this curve will be Kx_s. The ratio of the two slopes is therefore $Kx/Kx_s = x/x_s$. If the standard mixture consists simply of the pure absorber, then $x_s = 1$ and the ratio of the slopes gives the unknown concentration x.

In the above procedure, microwave power must be increased until the signal is maximum. For sample pressures of 50 to 100 mTorr, many microwave sources do not have sufficient output power to reach saturation and maximum signal. Presently available BWOs typically have maximum power outputs of 10 mW. In such cases the pressure may have to be lowered to 5 or 10 mTorr. This could be inconvenient in some cases. Signals may be smaller and pressure measurements are more difficult in this region.

An investigation by Funkhouser *et al.* (1968) indicates that concentrations can be determined accurately even if maximum signal is not obtainable. More study is needed, however, to ascertain whether the Γ method actually can be used for quantitative analysis in this circumstance (see Fig. 16).

Determinations of concentrations based on the Γ coefficient offer an added advantage. Implicit in the theory is a technique for detecting systematic errors of certain types such as sample changes during the measurements, overlapping absorption lines, or overlapping Stark lobes. This has been discussed by Harrington (1968).

2.2.4 *Accuracy and Precision*

In many cases the most severe limitation on the precision of a quantitative analysis by microwave spectrometry is likely to be imposed by the strength, or rather, the weakness of the absorption line. There has not been a sufficient amount of practical work reported in the literature to make possible reliable estimates of precision for various signal-to-noise ratios. A discussion of precision is therefore limited to estimates of the best precisions obtainable by the various methods assuming "favorable cases." One might guess that such cases would involve absorption lines with signal-to-noise ratios of 20 to 1 or better.

Concentrations can be determined with precisions of approximately 1% using relative intensity measurements provided the mixture is such that $\Delta\nu/p$ can be correctly assumed to be constant. In the more general case in which $\Delta\nu/p$ must be measured, the method of relative intensities is likely to give accuracies of approximately 5% because of limitations in measuring $\Delta\nu$. In very "favorable cases," however, $\Delta\nu$ and thus the concentration can, with care, be determined to approximately 1%.

Until the appearance of commercial spectrometers, the best measurements of the absolute value of the coefficient α_0 quoted precisions of 2 or 3% (Dymanus *et al.*, 1960). The Hewlett Packard Model 8460 will probably provide accuracies of approximately 2% in the measurement of α_0. The precision of the concentrations is then limited by measurements of $\Delta\nu$ and ranges from 1 to 5%.

Although reports in the literature are scarce, determinations of concentrations using the Γ theory seem to offer better results. No estimates are available for the procedure using a calibration curve formed by plotting the maximum signal against the partial pressure of the absorbing species. Precisions of 0.1% (Rinehart, 1969) or better were reported using the slopes of the plots of maximum relative signal versus total sample pressure. The precision of the method must vary considerably depending on the care taken in obtaining the slopes. This could be done quickly and roughly with only two measurements, or it could involve many data points and a least-squares fit.

Aside from experimental difficulties in obtaining precise measurements of α_0, $\Delta\nu$, signal, or pressure, there are possible sources of systematic errors that must be considered. In the Γ methods, for example, the crystal current should remain constant. Power saturation must be avoided if the procedure is based on the α_0 coefficient. If relative intensity measurements are used and concentrations are determined based on the α_0 coefficient, care must be taken that the power levels at the two transitions are equal. Incomplete or nonzero-based Stark modulation will lead to errors in any of the procedures outlined. For accurate measurements of absolute values of α_0, the Stark voltage must be sufficiently large to completely separate the Stark components from the main line. Errors in concentrations due to incomplete modulation can be minimized by using the same Stark field and pressure for the mixture and reference samples. Neighboring lines can cause overlapping problems and result in erroneous results (see Esbitt and Wilson, 1963). In an analysis, lines free of interference from nearby lines should be selected where possible. Reflections of microwave energy in the cell give spurious results owing to the fact that the energy passes through the sample more than once, thus altering the radiation path length. Reflections can be minimized by careful design and presumably are of secondary importance in commercial spectrometers in proper working order (Harrington, 1968).

The problem of selective adsorption of the sample on the cell walls is serious in quantitative analysis. This effect is especially large in Stark cells which use a large surface area of teflon to insulate the Stark septum. The effect can be minimized by "conditioning" the cell. In this process, successive samples are introduced into the cell until the concentration is no longer observed to vary with time. Heating the cell and continual pumping help in removing adsorbed molecules.

2.2.5 *Time Required for Analysis*

Because of the dearth of practical experience with microwave spectrometry in quantitative analysis, estimates of time required must be even more tenuous than estimates of accuracy. The most rapid technique, when it is applicable, is the use of relative intensities with $\Delta\nu/p$ constant. The time required for the analysis is essentially just that of sweeping through two lines. This would require only a few minutes to complete. In the isotopic determinations of Southern *et al.* (1952), the ^{15}N analysis, which included 10 to 15 measurements of the intensity ratios on a recorder, was performed in 20 to 30 min. The ^{13}C analysis, comprising 10 measurements of peak height ratios, took 50 min. Absolute measurements of α_0, line width, and pressure would require slightly more time, depending on the care taken. The above times do not include the time required to obtain the calibration curves.

If the approaches based on the Γ coefficient are used, several hours might be required if the calibration curves are plotted with care or a least-squares analysis is used to determine the slopes of the straight lines. If low precision is acceptable, one can draw the calibration plot by measuring only a few data points, and the process is speeded up considerably. Similarly, considerable time will be saved in cases where calibration curves, once determined, can be used repeatedly. The use of Eq. (30) would provide concentrations in a matter of a few minutes. It would first have to be ascertained that the error inherent in this approach is acceptable.

2.3 Literature Examples of Applications

The detection of industrial gaseous contaminants in the atmosphere is a problem for which microwave spectroscopy could provide a simple and reliable technique. In Fig. 10 we illustrate the qualitative identification of SO_2 in the presence of two other sulfur containing compounds. The assignment of the absorption lines to the various compounds was made by employment of the spectra obtained from pure samples. The strongest lines of each of the components of the gas mixture are used for the analysis. A concentration of a few parts per million of SO_2 should be readily detectable.

The sensitivity of low-resolution spectra to molecular composition is illustrated in Figs. 11–13. The rotational isomers of crotonic acid are clearly evident in the spectra of Fig. 11, where two sets of band spectra are found. Each absorption peak consists of a number of closely spaced lines which appear as a single unresolved line. Note the constant frequency separation between the weaker lines or between the stronger lines. Comparison of Figs. 12 and 13 shows that these chemical isomers give distinct spectra and are therefore easily distinguished.

Successful applications of microwave spectroscopy to the qualitative identification of the polar constituents of cigarette smoke and car exhaust have been reported by Harris and Jones (1971). Microwave spectroscopy has been applied by Kim and Gwinn (1964) to study the rearrangement of methylene groups in the reaction between cyclobutanol and PCl_5. Using a Stark cell coated with Corning K and a continuous flow system to reduce decomposition and reaction of radicals, Saito (1969) has confirmed the existence of the free radical SO as a product in the thermal decomposition of ethylene episulfoxide.

Very early in the development of microwave spectrometry the possibilities were recognized for the determination of abundance ratios of isotopes. The sensitivity of microwave spectra to isotopic composition is shown in Fig. 14. Townes *et al.* (1947) using relative intensity measurements of absorption lines of $^{35}ClCN$ and $^{37}ClCN$ determined the ratio of ^{37}Cl to ^{35}Cl to be 0.3. Wu *et al.* (1949) measured the concentration of the radioactive isotope ^{36}Cl.

Southern *et al.* (1952) developed a microwave procedure for isotopic

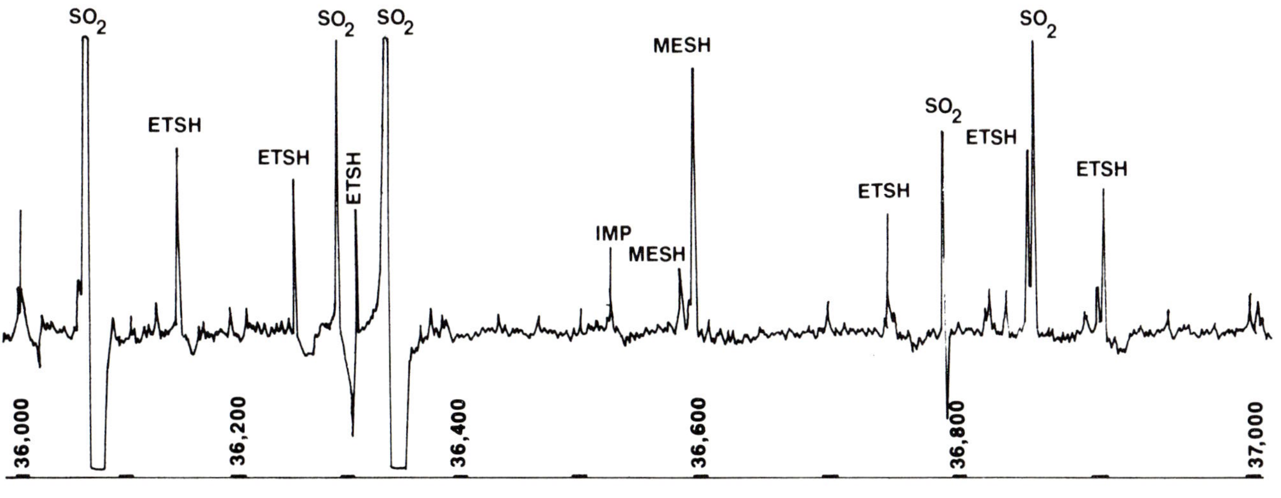

FIG. 10. Rotational spectrum of a gas mixture made up of 45% CH_3SH (MeSH), 45% CH_3CH_2SH (EtSH), and 10% SO_2. The absorption lines are up while the Stark lobes are shown downward. The presence of SO_2 is clearly indicated. The spectrum was run with a total gas pressure of 50 mTorr (volume is 500 cc), time constant of 0.3 sec, and a scan time of 1000 sec. [Used with permission of Hewlett Packard, Palo Alto, California.]

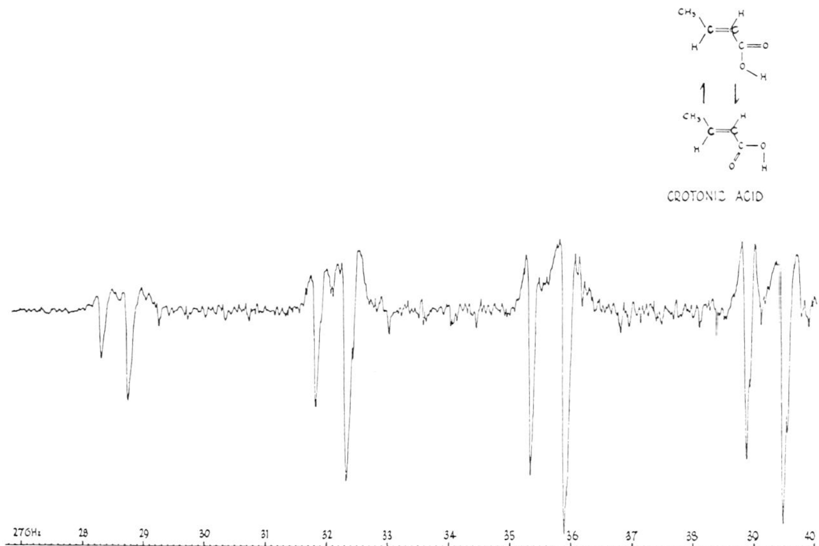

FIG. 11. Low-resolution microwave spectrum of crotonic acid. Scan rate 10 MHz/sec, pressure about 15 mTorr, and 1000-V Stark modulation. Spectra of two rotational isomers (*s*-trans and *s*-cis) are evident. Strongest bands are due to *s*-trans. Taken on a HP 8460A spectrometer.

analysis and used it to determine the concentrations of ^{15}N in ammonia and ^{13}C in cyanogen chloride. Adsorption of ammonia on the cell walls caused significant variations in pressure immediately after the sample entered the cell. Measurements were performed only after the pressure became constant. Pressures as high as 100 mTorr were used to minimize this effect. To remove the adsorbed gas, the cell was heated with continual pumping. It is expected that the line of the parent and the isotopic species have equal widths at a given pressure so that the ratio of peak intensities is the same as that of integrated intensities. Peak heights were measured from the peak of the main line to the peak of the unresolved Stark component. This eliminated uncertainties in the position of the base line. Because of nonlinearities in the spectrometer system, calibration curves were used. Rather than plotting ratios of peak heights of $^{15}NH_3$ to $^{14}NH_3$ versus concentration of $^{15}NH_3$, calibration curves were made by first measuring these ratios for various prepared concentrations of $^{15}NH_3$ and then dividing the ratios by the ratio of peak heights of $^{15}NH_3$ to $^{14}NH_3$ in normal ammonia. This was plotted against concentrations of $^{15}NH_3$. This procedure largely

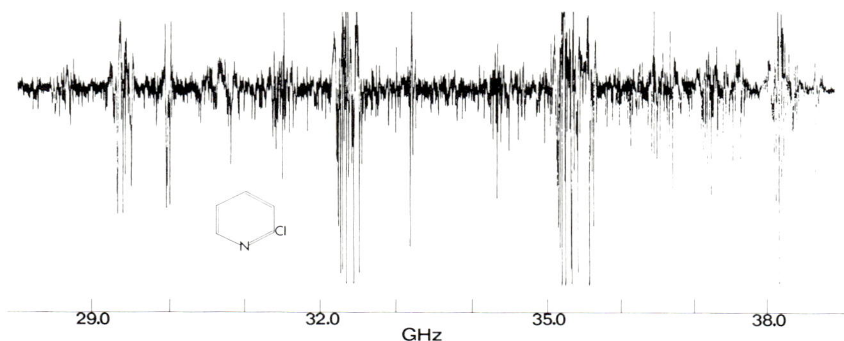

FIG. 12. Low-resolution R-band spectrum of 2-chloropyridine. Scan rate about 3 MHz/sec, time constant 0.4 sec, and 130-V Stark modulation. The ^{37}Cl species give a cluster of lines at lower frequency from the strong ^{35}Cl lines. (See R. T. Walden, 1973.)

avoids errors due to variations of peak ratios that might occur as a result of changes in spectrometer characteristics, since variations caused the enriched and normal ratios to change proportionately. By measuring the $^{15}N/^{14}N$ or $^{13}C/^{12}C$ ratio in an unknown sample, the concentration of ^{15}N or ^{13}C can be obtained from the calibration curves. Typical spectra obtained for the normal ($^{15}NH_3$ = 0.38%) and enriched samples of ammonia are illustrated in Fig. 15. For ^{15}N, occurring in the range of 0.38 to 4.5%, accuracy was within 3% of its concentration. For ^{13}C, in the range of 1.1 to 10%, the average error was less than 2% of its concentration. An analysis required approximately 0.00015 moles of gas, most of which was recoverable.

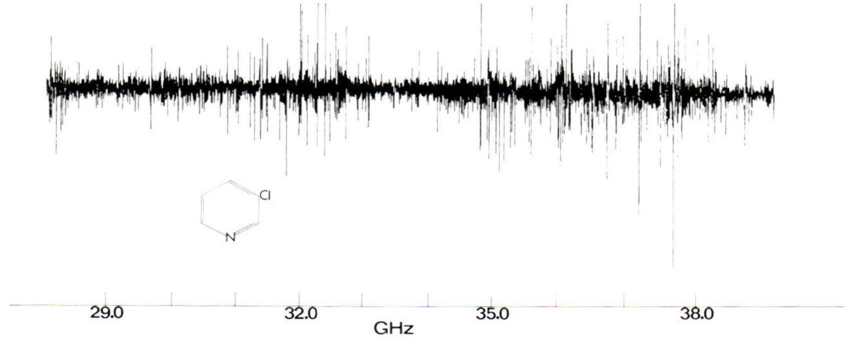

FIG. 13. Low-resolution R-band spectrum of 3-chloropyridine. Scan rate about 3 MHz/sec, time constant 0.4 sec, and 130-V Stark modulation. Sample obtained from Aldrich Chemical Co., Inc., Milwaukee, Wis., and used without further purification.

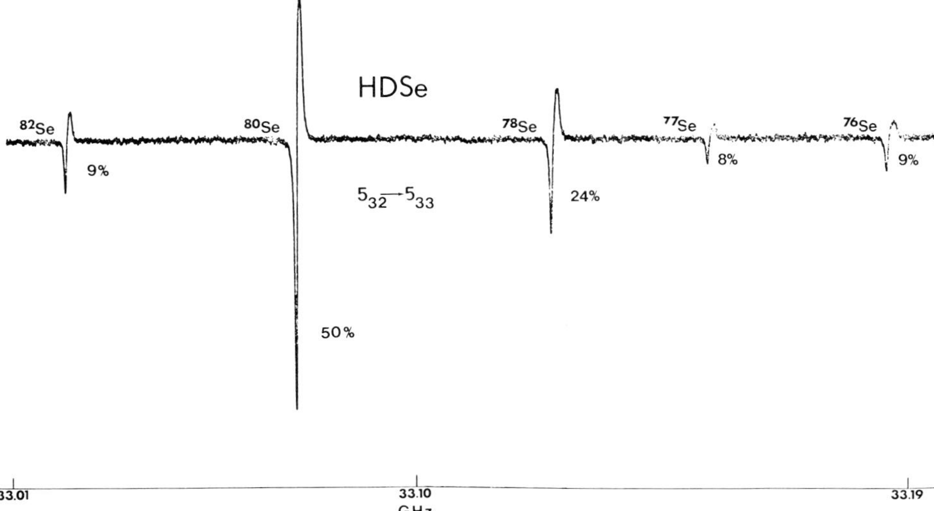

FIG. 14. A transition of the asymmetric rotor HDSe. Spectral trace recorded from left to right and extends approximately 170 MHz. The five major isotopic species of Se are clearly evident. The spectrum was taken at the temperature of dry ice; nevertheless, HDSe decomposes in the Stark cell. This is evident from the decrease in line intensity of ^{76}Se compared with that of ^{82}Se.

Although these results are quite good, one would expect that the above procedure would yield much better results on the commercial spectrometers now available.

Weber and Laidler (1951) used microwave spectroscopy to determine concentrations of NH_3 in mixtures of deuterated ammonias and to study the kinetics of the catalyzed NH_3–D_2 isotopic exchange reaction. An unmodulated spectrometer was used, and concentrations were determined by merely comparing the detector signals with and without the presence of the sample. The poor sensitivity inherent in this technique is sufficient to detect only extremely strong absorbers. The ammonia concentrations determined were considered accurate to within 0.5%.

Esbitt and Wilson (1963) have measured peak intensity ratios with accuracies of 2 or 3%. This was accomplished using a conventional Stark spectrometer with simple modifications. Considerable care was taken to remove several sources of error, including multiple reflections in the waveguide. Errors due to nonlinear response of the detector and amplifier were avoided by reducing the peak height of the larger of a pair of lines with a calibrated attenuator until it nearly equaled the height of the smaller line.

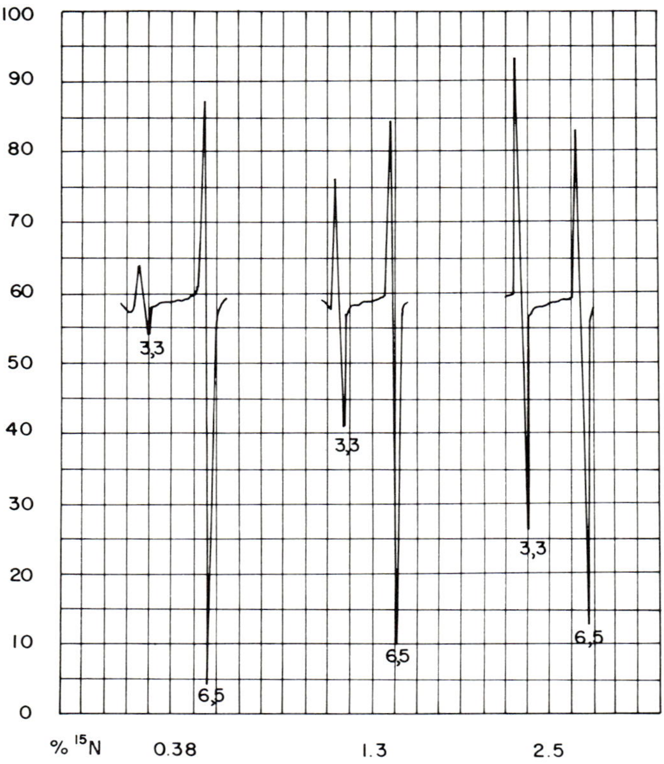

FIG. 15. Typical spectra obtained with normal and ^{15}N enriched samples of ammonia at about 125 mTorr pressure. The (6,5) $^{14}NH_3$ line (22,732.43 MHz) is recorded with less amplifier gain (factor of 10) than the (3,3) $^{15}NH_3$ line (22,789.41 MHz). The absorption lines are shown downward and the Stark components are upward. [After A. L. Southern, H. W. Morgan, G. W. Keilholtz, and W. V. Smith (1951). *Anal. Chem.* **23**, 1000. Copyright (1951) by the American Chemical Society. Used with permission.]

In the unsaturated region, the ratio of the signals produced by two different absorption lines is equal to the ratio of their $\alpha_0 l$ provided the power level, or, equivalently, the rectified crystal current is held constant. For nearby lines, the error introduced by different effective path lengths l caused by multiple reflections was found to be small in their spectrometer. The ratio of the abundance of the ^{12}C to the ^{13}C species of enriched methyl formate was determined to be 0.635 with a 2.8% standard deviation by using measurements on the same transition in the two isotopic species and taking C in Eq. (24) to be constant. This was in good agreement with a mass spectroscopic determination which gave 0.644 with a 3% uncertainty.

Rinehart (1969) used the method of slopes based on the Γ coefficient (described in Section 2.2.3) to determine concentrations of methanol in benzene with accuracies of approximately 0.1%. Great care was taken and a least-squares fit was used to determine the slopes of the curves.

Funkhouser *et al.* (1968) have recently used microwave spectrometry to determine the concentrations of acetone and freon in nitrogen. In one sample, the concentrations of acetone and freon were measured to be 0.2 and 0.9%, respectively. The results were in good agreement with the mass spectrometry values 0.2 and 1.0%. Operating with power levels such that the signal is only near maximum, the investigators found that a plot of output signal versus sample pressure for a mixture of acetone in nitrogen served as an accurate calibration curve for another acetone–nitrogen mixture (see Fig. 16).

Hirota and co-workers (1964, 1966) and Hironaka *et al.* (1966) used

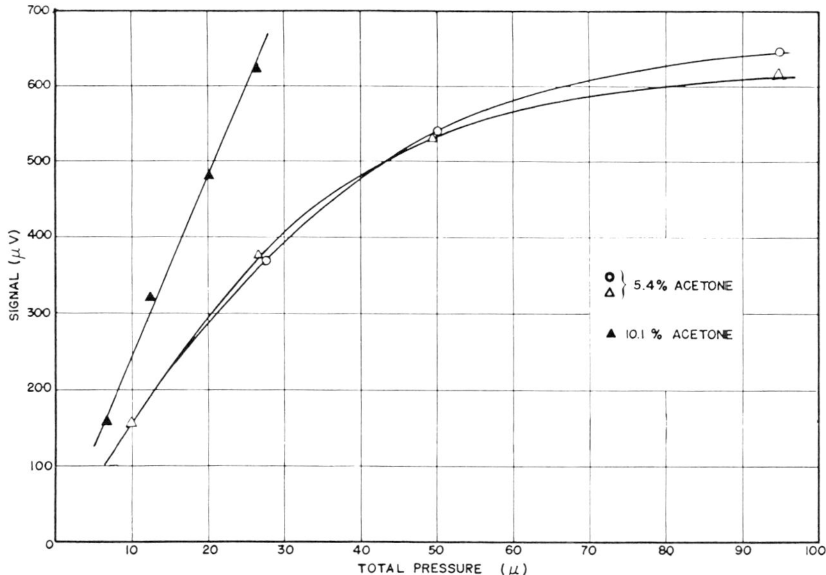

FIG. 16. Calibration data of acetone in nitrogen. Source power at 0.5 mW; ○ and △ indicate 5.4% acetone, ▲ indicates 10.1% acetone. Above about 20 μm nonlinear curves are obtained for the 5.4% mixture, since sufficient power to reach maximum signal was not available. Note that a signal of 400 μV corresponds to a partial pressure of acetone equal to 1.6 mTorr in both the 5.4% mixture (5.4% × 30 μm = 1.6 μm) and the 10.1% mixture (10.1% × 16 μm = 1.6 μm). [After Funkhouser *et al.* (1968) *Anal. Chem* **40,** 22A.]

relative intensities of microwave absorption lines to measure the relative amounts of products obtained in the isotopic exchange of propene with deuterium oxide in the presence of nickel, palladium, and platinum catalysts.

Aleksandrov and Tysovskii (1967), using a flow system and a broad-banded microwave spectrometer, measured relative intensities to determine the concentration of ethanol and isopropanol in gasoline. An average experimental error of 10% was obtained. The success of microwaves in this determination is significant, as analysis of this mixture by other methods is difficult.

Sakurai *et al.* (1971) measured relative intensities to determine the abundances of the isotopic species resulting from replacing hydrogen by deuterium in the isomerization of *cis*-2-butene and 1-butene.

Scharpen *et al.* (1972) performed a very careful isotopic analysis of the products of the hydrogen–deuterium exchange between CH_3OD and propene-d_0 in the presence of platinum, rhodium, and nickel. Errors of approximately 2% were obtained in the concentrations, which ranged from 1.5 to 17% of the total sample. An interesting aspect of this investigation was the use of a computer to control the spectrometer frequency, the modulation voltage, and also to perform signal averaging and data reduction. Analytical applications of microwave spectrometry in conjunction with a computer have also been discussed by White (1970).

Zeman and Zemanova (1972) used microwave spectrometry to determine moisture contents in a stage in the production of toilet soap. Water concentrations as low as 0.15% were determined.

Crable (1972) measured small concentrations of air pollutants by using both relative intensity techniques and integrated line intensities. Although the method of relative intensities proved most convenient and accurate, integrated line intensities, measured with a planimeter, gave accuracies of a few percent. Procedures based on the Γ coefficient were inapplicable because of the large amount of microwave power necessary to obtain maximum signal. Crable found that a conventional microwave spectrometer was not as sensitive for this application as mass spectrometry or gas chromatography. This suggests the use of a microwave cavity spectrometer rather than a conventional spectrometer for such problems.

Using a simplified cavity spectrometer, Hrubesh (1971) performed qualitative analyses on polluted air samples. Preliminary measurements indicated a sensitivity capable of detecting 11 ppm for NO_2 and 3 ppm of SO_2 without sample enrichment. It was felt that the performance of the spectrometer could be significantly improved. The construction and use of such a system for monitoring specific trace pollutants is discussed.

Srinivasan (1973) very recently reported on the possibility of the application of microwave spectrometry to determinations of ammonia concentrations in blood serum. A Stark-modulated cavity spectrometer was used, so that sufficient power density was available to make use of techniques based on the Γ coefficient. Srinivasan used a modulation configuration that resulted in less than optimum sensitivity, however, and minimum concentrations of NH_3 in MeOH of no better than 100 ppm were detected.

References

Aleksandrov, A. N., and Tysovskii, G. I. (1967). *Zh. Anal. Khim.* **22,** 1286.

Baird, D. H., and Bird, G. R. (1954). *Rev. Sci. Instrum.* **25,** 319.

Beers, Y. (1959). *Rev. Sci. Instrum.* **30,** 9.

Bird, G. R. (1954). *Rev. Sci. Instrum.* **25,** 324.

Bleaney, B., and Penrose, R. P. (1947). *Proc. Phys. Soc. (London)* **60,** 83.

Cord, M. S., Lojko, M. S., and Petersen, J. D. (1968). "Microwave Spectral Tables," Vol. 5. NBS Monograph 70, U. S. Dep. of Commerce, Washington, D.C.

Crable, G. F. (1972). U.S. Nat. Tech. Inf. Serv., PB Rep. No. 212554.

Crable, G. F., and Wahr, J. C. (1969). *J. Chem. Phys.* **51,** 5181.

Curl, Jr., R. F. (1969). *J. Mol. Spectrosc.* **29,** 375.

Cuthbert, J., Denney, E. J., Silk, C., Stratford, R., Farren, J., Jones, T. L., Pooley, D., Webster, R. K., and Wells, F. H. (1971). "The Design of an Analytical Microwave Spectrometer." Cambridge Scientific Instruments, Cambridge, England.

Danos, M., and Geschwind, S. (1953). *Phys. Rev.* **91,** 1159.

Dymanus, A. (1959). *Rev. Sci. Instrum.* **30,** 191.

Dymanus, A., Dijkerman, H. A., and Zydervield, G. R. D. (1960). *J. Chem. Phys.* **32,** 717.

Esbitt, A. S., and Wilson, E. B. (1963). *Rev. Sci. Instrum.* **34,** 901.

Funkhouser, J. T., Armstrong, S., and Harrington, H. W. (1968). *Anal. Chem.* **40,** 22A.

Gordy, W. (1956). *In* "Chemical Applications of Spectroscopy," pp. 71–185. Wiley (Interscience), New York.

Gordy, W. (1965). *In* "Pure and Applied Chemistry," Vol. 2, pp. 403–434. Butterworth, London and Washington, D.C.

Gordy, W., and Cook, R. L. (1970). "Microwave Molecular Spectra," Part II of Chemical Applications of Spectroscopy, 2nd ed. Wiley, New York.

Gordy, W., Smith, W. V., and Trambarulo, R. F. (1953). "Microwave Spectroscopy." Wiley, New York. Republication, 1966, Dover, New York.

Hardy, W. A., Fletcher, P., and Suarez, V. (1954). *Rev. Sci. Instrum.* **25,** 1135.

Harrington, H. W. (1967). *J. Chem. Phys.* **46,** 3698.

Harrington, H. W. (1968). *J. Chem. Phys.* **49,** 3023.

Harris, C. A., and Jones, G. E. (1971). Mississippi Acad. Sci. Meeting.

Helminger, P., De Lucia, F. C., and Gordy, W. (1970). *Phys. Rev. Lett.* **25,** 1397.

Hironaka, Y., Hirota, K., and Hirota, E. (1966). *Tetrahedron Lett.* **22,** 2437.

Hirota, K., and Hironaka, Y. (1966). *Bull. Chem. Soc. Japan* **39,** 2638.

Hirota, K., Hironaka, Y., and Hirota, E. (1964). *Tetrahedron Lett.* **25,** 1645.

Hrubesh, L. W. (1971). UCRL 73197, presented at Joint Conf. on Measurement of Environmental Pollutants, Palo Alto, California.

Hughes, R. H. (1952). *Ann. N. Y. Acad. Sci.* **55**, 872.
Jones, G. E., and Beers, E. T. (1971). *Anal. Chem.* **43**, 656.
Kewley, R., Sastry, K. V. L. N., Winnewisser, M., and Gordy, W. (1963). *J. Chem. Phys.* **39**, 2856.
Kim, H., and Gwinn, W. D. (1964). *Tetrahedron Lett.* **37**, 2535.
Lichtenstein, M., Gallagher, J. J., and Cupp, R. E. (1963). *Rev. Sci. Instrum.* **34**, 843.
Lide, Jr., D. R. (1966). *Advan. Anal. Chem. Instrum.* **5**, 235–277.
McAfee, K. B., Hughes, R. H., and Wilson, E. B. (1949). *Rev. Sci. Instrum.* **20**, 821.
Mattuck, R. D., and Strandberg, M. W. P. (1958). *Rev. Sci. Instrum.* **29**, 717.
Rinehart, E. A. (1969). NASA CR-107171.
Rinehart, E. A., Legan, R. L., and Lin, C. C. (1965). *Rev. Sci. Instrum.* **36**, 511.
Saito, S. (1969). *Bull. Chem. Soc. Japan* **42**, 667.
Sakurai, Y., Takaharu, O., and Tamaru, K. (1971). *Trans. Faraday Soc.* **67**, 3094.
Scharpen, L. H., and Laurie, V. W. (1972). *Anal. Chem.* **44**, 378R.
Scharpen, L. H., Rauskolb, R. F., and Tolman, C. A. (1972). *Anal. Chem.* **44**, 2010.
Sheridan, J. (1973). *In* "MTP International Review of Science," Vol. 12, pp. 251–277. University Park Press, Baltimore, Maryland.
Southern, A. L., Morgan, H. W., Keilholtz, G. W., and Smith, W. V. (1951). *Anal. Chem.* **23**, 1000.
Srinivasan, T. M. (1973). *Phys. Scripta* **7**, 84.
Sugden, T. M., and Kenney, C. N. (1965). "Microwave Spectroscopy of Gases." Van Nostrand Reinhold, Princeton, New Jersey.
Townes, C. H., and Schawlow, A. L. (1955). "Microwave Spectroscopy." McGraw-Hill, New York.
Townes, C. H., Holden, A. N., and Merritt, F. R. (1947). *Phys. Rev.* **71**, 64.
Verdier, P. H. (1958). *Rev. Sci. Instrum.* **29**, 646.
Verdier, P. H., and Wilson, E. B. (1958). *J. Chem. Phys.* **29**, 340.
Wacker, P. F., and Pratto, M. R. (1964). "Microwave Spectral Tables," Vol. 2. NBS Monograph 70, U. S. Dep. of Commerce, Washington, D. C.
Walden, R. T. (1973) Ph.D. thesis. Mississippi State Univ., Mississippi State, Mississippi.
Weber, J., and Laidler, K. J. (1951). *J. Chem. Phys.* **19**, 1089.
White, W. F. (1970). "Automation and the Microwave Spectrometer," presented at the Society for Applied Spectroscopy Meeting, New Orleans, Louisiana.
Wollrab, J. B. (1967). "Rotational Spectra and Molecular Structure." Academic Press, New York.
Wu, C. S., Townes, C. H., and Feldman, L. (1949). *Phys. Rev.* **76**, 692.
Zeman, I., and Zemanova, D. (1972). *Veda Vyzk. Potravin Prum.* **23**, 247.

CHAPTER 12

Neutron Activation Analysis

H. R. Lukens,* H. L. Schlesinger, and D. Bryan

Gulf Radiation Technology
San Diego, California

Introduction

Neutron activation analysis (NAA) is a method for determining how much of various chemical elements is present in a sample. The method utilizes neutrons to generate radioactive analytical indicator isotopes, and the radiosiotopes are then measured by means of the radiations that they emit. The production of the indicator radioisotopes, called "activation," involves nuclear reactions; also, the emission of radiation from the indicators derives from nuclear processes. Hence, the method is intrinsically independent of chemical form, and in fact, it is not useful for determining the chemical compounds in which various elements reside.

Each of several possible nuclear reactions is carried out on any given element during the activation step, and no two elements give rise to the same resulting array of product isotopes. Since each product radicisotope

* Present address: Intelcom Rad Tech, San Diego, California.

has its own unique decay characteristics, NAA provides qualitative analyses. Furthermore, all else being equal, the amount of an indicator produced by activation is directly proportional to both (1) the amount of the element from which it is formed in the sample, and (2) the intensity of its emitted radiation. Thus, the method is also quantitative.

1 Theory

1.1 Activation Theory

In the process of activation, neutrons are spoken of as bombarding particles, and the nuclei that they interact with are called target nuclei. Target nuclei and the bombarding neutrons that hit them undergo some sort of reaction. Sometimes the reaction is simply one of scattering (something like billard balls), but there is a good chance that the neutron will interact more intimately with the nucleus.

Neutrons are bound to nuclei with millions of electron volts (usually 6–8 MeV) of energy—i.e., about that amount of energy is required to remove a neutron from a nucleus. Thus, if a thermal (0.025 eV) neutron is captured by a nucleus, the resulting compound nucleus will have a large excess of energy. This excess energy is promptly thrown off as gamma rays (in all but a minority of cases wherein the compound nucleus may split in two), and a product isotope of the target nucleus results. The reaction may be written ${}^{m}A + \mathrm{n} = {}^{m+1}A + \gamma$, where A is the atomic number of the reaction nucleus and m is its number of nucleons. For convenience, the reaction is usually written in abbreviated form:

$${}^{m}A\,(\mathrm{n}, \gamma)\,{}^{m+1}A,$$

and this type of reaction is called an (n, γ) reaction.

If the bombarding neutron has enough energy, it can cause reactions that involve the ejection of particles from the target nucleus. The more common reactions of this type are

(a) (n, 2n) reaction: i.e., ${}^{m}A + \mathrm{n} = {}^{m-1}A + 2\mathrm{n}$,
(b) (n, p) reaction: i.e., ${}^{m}A + \mathrm{n} = {}^{m}(A - 1) + \mathrm{p}$, where p is a proton, and
(c) (n, α) reaction: i.e., ${}^{m}A + \mathrm{n} = {}^{m-3}(A - 2) + \alpha$, where α is an alpha particle.

Under certain circumstances, as in a nuclear reactor having a wide spectrum of neutron energies, all of these possible reactions can occur. For example, taking natural fluorine, ${}^{19}\mathrm{F}$, as target nuclei, the reactions taking place will be as follows:

^{19}F (n, γ) ^{20}F, ^{19}F (n, p) ^{19}O, ^{19}F (n, α) ^{16}N, and ^{19}F (n, 2n) ^{18}F.

All of the foregoing product species are radioactive isotopes and any or all may be useful indicator isotopes.

The reaction rate P for a specified reaction is a function of the number of target atoms N, the flux of neutrons of specified energy ϕ, and the cross section σ, of a target nucleus toward those neutrons for that particular reaction:

$$P = N\phi\sigma \tag{1}$$

Radioisotopes decay at the disintegration rate described by the first-order equation

$$-dN^*/dt = \lambda N^* \tag{2}$$

where N^* is the number of atoms of the specified isotope and λ is the decay constant of that isotope. Usually, the half-life $t_{1/2}$ values for radioisotopes are given in the literature; λ may be computed from $t_{1/2}$ by the relationship $\lambda = 0.693/t_{1/2}$.

Of course, radioisotopes formed during an irradiation are subject to decay as soon as they are formed. Hence the net production rate of a given isotope combines Eqs. (1) and (2):

$$dN^*/dt = P - \lambda N^* \tag{3}$$

Integration of Eq. (3) gives the equation

$$N^*\lambda = P(1 - e^{-\lambda t_i}) \tag{4}$$

where t_i is the duration of the irradiation. Equation (4) describes the disintegration rate of a given isotope at the end of the irradiation. The disintegration rate is often called the "activity," and is denoted by A. A_0 usually indicates the activity of the isotope at the end of the irradiation. Also, $(1 - e^{-\lambda t_i})$ is called the saturation term s. Thus, the equation often used in activation analysis, and called the "activation equation," is

$$A_0 = N\phi\sigma s \tag{5}$$

1.2 Measurement and Identification of Analytical Indicator Radioisotopes—Gamma-Ray Spectrometry

Subsequent to the end of the irradiation the activity of a given radioisotope may be computed by the integral of Eq. (2) multiplied by λ:

$$A = A_0 e^{-\lambda t_d} \tag{6}$$

where t_d is the decay period subsequent to the end of irradiation. Obviously, by measuring A at two different postirradiation decay periods, the decay constant can be obtained. This is sometimes used as a means of identifying the radioisotope being measured. However, since essentially all elements in a sample are subject to activation (and in most cases more than one radioisotope is produced from an element), an activated sample is usually a complex mixture of isotopes, and additional means of identification are needed.

At one time radiochemistry was the main tool for separating the radioisotopes of one element from those of the other elements. The amount of the isolated radioisotope was measured with a Geiger counter and its radiopurity verified by making successive measurements and calculating the apparent half-life. However, in the late 1950s the thallium-activated sodium iodide scintillation counter and electronic pulse height analyzers had been developed to a point where fairly complex mixtures of gamma ray emitting radioisotopes could be resolved by gamma ray spectrometry. Since most of the indicator radioisotopes used in NAA emit gamma rays, this development greatly improved the speed and economy of NAA. Even where the mixture was too complex to be resolved by gamma ray spectrometry, the tool allowed greatly reduced effort in radiochemical processing.

More recently detectors with much greater resolving power than the NaI(Tl) have been developed, the most useful (for NAA) of which is lithium-drifted germanium, Ge(Li). The resolving powers of NaI(Tl) and Ge(Li) detectors for 662 keV gamma rays are approximately 8% and 0.5%, respectively. Thus, the latter detector can usefully examine much more complex mixtures than the former. However, Ge(Li) detectors are comparatively small and expensive, which leads to higher cost per sample measured. For these reasons, use of the NaI(Tl) counter is indicated where sample complexity is not excessive.

Both kinds of γ-ray detectors provide electrical signals that are linearly proportional to the amount of γ-ray energy that they absorb. Gamma-ray spectra display peaks corresponding to complete absorption of γ rays of each particular energy. Gamma rays ≥ 1.022 MeV can cause the production of positron–electron pairs, and positrons will interact with electrons to produce two 511-keV γ rays. When this happens in the detector, one or both of the 511-keV γ rays may escape, and "escape" peaks corresponding to 511 and 1.022 MeV less than the full γ-ray energy can be observed. Incomplete absorption of γ rays in the detector, due to Compton scattering, gives rise to a continuum of "noise" in the spectrum at energies below peak energy. Other characteristics of spectra are 180° backscatter peaks, x rays stimulated by γ-ray interaction processes, and "sum" peaks. The "sum" peaks correspond to the complete absorption in the detector of two γ rays at the same time. Thus, while the information contained in a γ-ray spectrum is highly meaningful, care must be taken in its interpretation.

For NAA purposes the peaks that correspond to complete γ-ray absorption, and their intensities, are of primary value. Identification of peaks and estimation of their intensities are fairly simple, provided the spectrometer has been calibrated with γ rays of known energy and intensity, as described in a later section.

Although an isotope may give rise to a particular γ ray, it may not do so in every disintegration. The yield Y of the various γ rays from isotopes is given by Lederer *et al.* (1967). Furthermore, the combined geometric and response efficiency E is such that only a fraction of emitted γ rays of a particular energy will be recorded in the peak for that energy (the factor E may be determined by calibration with reference standards). Therefore, if we wish to compute a peak's count rate C, the activation equation, Eq. (5), must be appropriately modified:

$$C_0 = N\phi\sigma sYE \tag{7}$$

In practice the analyst usually uses Eq. (7) in designing his experiments. Precision determinations are then made with the use of comparator standards, as explained in a later section.

The number N is obtained from the estimated weight of the element in the sample and Avogadro's number. The value of ϕ is usually defined by the operators of the device that supplies neutrons (nuclear reactors or machine accelerators), although it may be desirable to carry out independent flux calibrations in some cases. The above source of Y values may also be used to obtain the half-life values needed to compute s; E is calibrated with standardized reference radioisotopes; and values for σ for thermal neutron (n, γ) reactions are found in Hughes and Schwartz (1958). Values for reactor fission-spectrum neutron cross sections are given in Roy and Hawton (1960). Kenna and Conrad (1966) have listed 14-MeV neutron reaction cross sections, together with other data of use to analysts who wish to use machine generators that produce such neutrons.

2 Applications and Limitations

2.1 Qualitative and Quantitative Uses

NAA is rarely employed as a qualitative analytical technique. Even when only qualitative information is required, the nature of the radiation detection equipment is such that the data needed for a quantitative determination are available.

2.1.1 *Selectivity*

Once a particular nuclear reaction has been chosen for use in an analytical procedure, three factors can be manipulated to enhance selectivity. These are the spectrum of neutron energies impinging on the sample, the duration of the various analytical steps (length of irradiation and delay before counting), and the radiation detection equipment employed.

Three neutron energy intervals are of importance to NAA. The lowest varies from 0 to 0.026 eV, and is known as the "thermal" region. Thermal neutrons generally are easily captured by atoms, and their capture favors the important (n, γ) reaction. They can be produced from higher energy neutrons by interposing a low atomic number element (especially hydrogen) between the source and sample. Epithermal neutrons range from 0.027 to about 100 eV. They are important because many elements have extremely high capture cross sections in narrow energy bands called "resonances" which occur in this region. Epithermal neutron reactions also are generally (n, γ). They are produced in the same way as thermal neutrons. In some cases it is possible to suppress interference from competing reactions by producing a flux rich in epithermal relative to thermal neutrons. This is accomplished by interposing a material with very high thermal neutron capture cross section (especially cadmium) between the source and sample. If the desired reaction has a large epithermal resonance, it will occur with good yield relative to competing reactions without such resonances.

Energies above 100 eV are associated with "fast" neutrons. These neutrons tend to induce (n, p), (n, 2n), and (n, α), known as "fast reactions." As the neutron energies increase above 1 MeV, fast reactions predominate and (n, γ) reactions tend to become rare. Fast neutrons are produced by all neutron sources usually employed in NAA.

Of the various neutron sources available, three are particularly useful in NAA. Two are the fission spectrum sources, i.e., nuclear reactors and ^{252}Cf; the third is the fast neutron source known as the neutron generator.

In nuclear reactors neutrons are produced as a by-product of the self-sustaining fission reaction. ^{252}Cf is an isotope which undergoes spontaneous fission and has recently become available in quantities sufficient for useful NAA applications. Both produce neutrons which range in energy from less than 0.1 to over 10 MeV. The modal neutron energy is about 1 MeV at the source, but can easily be thermalized with hydrogenous material. In water-moderated nuclear reactors the ratio of thermal to fast plus epithermal neutrons at sample positions is approximately unity.

The operation of the neutron generator is discussed below in the analysis of oxygen in beryllium.

Time can be used to increase the selectivity of NAA. Long irradiations saturate the amount of short half-lived activity in the sample, while continuing to increase the amounts of longer lived activities. Short irradiations thus favor the production of short half-lived nuclides. Nuclear reactors can be operated in a "pulsed" mode in which extremely high neutron fluxes are produced in a burst measured in milliseconds. Very short half-lived activities such as 0.8 sec $^{207}Pb^{m}$ can become sensitive analytical indicators under these conditions. Lukens *et al.* (1965) discuss this subject in detail. Repeated short irradiations and short counting periods, using the same sample, can be used to accumulate γ-ray spectra in which short-lived activities predominate.

The time between the end of the irradiation and the beginning of the count can also be used to increase selectivity. If a short-lived activity relative to the one to be determined is present in a sample, the beginning of counting is delayed until it decays to insignificance. If long half-lived activities relative to the one to be determined are present, the counting is begun before they predominate. In general, both types of unwanted activities are present and some optimum time is selected for beginning the counting operation.

The fundamentals of γ-ray spectrometry were discussed in Section 1.2 on theory. Some refinements are discussed under the topic of Sensitivity (Section 2.1.2).

While γ-ray spectrometry is an important tool to the activation analyst, it is not his only resource. One very selective analytical technique employs the BF_3-filled neutron detector to detect fissionable isotopes. The basis for the technique is the fact that some daughter isotopes of the "(n, f)" or neutron-induced nuclear fission reaction decay with the emission of neutrons.

Since neutron detectors are practically insensitive to β or γ rays, the technique can be used to advantage in detecting very low levels of fissionable isotopes in soil, water, or air samples. Another fertile source of samples for this technique is in the assay of nuclear fuels. The cross section of the (n, f) reaction is much higher for ^{235}U than for the other uranium isotopes, allowing this important isotope to be determined in samples of varying isotopic composition.

A few isotopes of interest in NAA do not emit γ rays, but decay by pure β process. These samples must be analyzed by β-ray counting, and generally require radiochemical treatment to transform them into a suitably pure state. ^{32}P is one such isotope. It serves as an analytical indicator for P via the $^{31}P(n, \gamma)^{32}P$ reaction and for S via the $^{32}S(n, p)^{32}P$. It is also produced by the (n, α) reaction of ^{35}Cl. When β counting is required, the half-life of the material should be verified by repetitive counting to insure its radiopurity.

At times it is not sufficient to identify the product isotope. Consideration must also be given to which of one or more competing reactions gave rise to the observed isotope. For example, ^{56}Mn can be produced by the $^{56}Fe(n, p)^{56}Mn$, the $^{55}Mn(n, \gamma)^{56}Mn$, or the $^{59}Co(n, \alpha)^{56}Mn$ reactions. Oftentimes even a modest amount of information about the sample and access to tables of reaction cross sections will serve to eliminate all but one possible reaction.

Table 1 gives some pertinent data from Filby *et al.* (1969). If the ratio of thermal to epithermal and fast neutrons is approximately unity at the sample irradiation position (the actual value should be determined experimentally for a particular system), it is apparent that ~27,000 μg of Fe or ~92,000 μg of Co are required to produce the same amount of ^{56}Mn activity as 1 μg of ^{55}Mn. If it is known that neither iron nor cobalt are major constituents of the sample, it can be deduced that only the (n, γ) reaction is of importance. This is an example of the general rule that cross sections for (n, γ) reactions are much larger than for other neutron-induced reactions.

If the iron and cobalt levels in the sample are completely unknown, standards of the elements can be irradiated. The yields of ^{56}Mn produced in these standards are determined as well as the yields of ^{59}Fe and ^{60}Co. If significant levels of ^{59}Fe or ^{60}Co are detected in the sample, appropriate corrections can be made. Once this exercise is accomplished for a particular

TABLE 1

CROSS SECTIONS AND ABUNDANCES FOR SOME NEUTRON-INDUCED REACTIONS

Parent isotope	Relative abundance (%)	Nuclear reaction	Product isotope	Cross section (barns[a])
^{55}Mn	100.	(n, γ)	^{50}Mn	13.
^{50}Fe	92.	(n, p)	^{56}Mn	0.00044
^{59}Co	100.	(n, α)	^{56}Mn	0.00014
^{58}Fe	0.31	(n, γ)	^{59}Fe	1.1
^{59}Co	100.	(n, γ)	^{60}Co	18.

[a] 1 barn = 10^{-24} cm^2.

element, irradiation position, and detector system, the yields of the various reactions can be tabulated and used in planning subsequent experiments. Lukens (1964) has published a catalog of such data, including many theoretically derived values for reactions whose yields are too small to be measured directly.

Occasionally no suitable independent indicator for the parent isotope of a competing reaction can be found. In such cases, the fact that reaction cross sections are functions of neutron energy can be used to provide an independent measurement. Suppose that silicon is to be determined in a sample that may contain trace levels of aluminum. It is decided to employ the $^{28}Si(n, p)^{28}Al$ reaction. Unfortunately, the $^{27}Al(n, \gamma)^{28}Al$ reaction is quite sensitive and no other neutron-induced reaction on aluminum is available which is sensitive enough to independently measure its level in the sample. Two identical aliquots of the sample are prepared, one of which is wrapped in cadmium foil. The (n, γ) reaction will be depressed in the aliquote wrapped in the thermal neutron absorbing foil, while the (n, p) reaction yield is almost unaffected. Aluminum and silicon standards are prepared in duplicate and one of each pair is foil wrapped. Samples and standards are irradiated and counted under identical conditions. The 1.78-MeV photopeak due to 2.3-min ^{28}Al is measured in each. Then

$$N_1 = S_1 + A_1 \tag{8}$$

where N_1 is the net corrected yield obtained from the bare sample, which is composed of a contribution S_1 from silicon and A_1 from aluminum. Similarly,

$$N_2 = S_2 + A_2 \tag{9}$$

for the wrapped sample. The ratio

$$f = S_1/S_2 \tag{10}$$

is computed as the net corrected yield for the bare silicon standard divided by that from the clad standard. Similarly,

$$g = A_1/A_2 \tag{11}$$

for the aluminum standards. Then

$$S_2 = (N_1 - gN_2)/(f - g) \tag{12}$$

and the silicon level in the sample can be determined by comparison with the clad silicon standard.

2.1.2 *Sensitivity*

Many factors at least partially under the control of the experimenter control the sensitivities attainable by NAA. The intensity and duration of the irradiation, the duration of the decay and counting periods, the size of the sample analyzed, the type and size of the detector employed as well as the relative positions of detector and sample will all strongly influence the sensitivity obtained. These have been discussed in Sections 1.1 and 2.1.

In Table 2 are given the limits of detection which are routinely attained at the authors' laboratory with irradiation and counting intervals not exceeding 1 hr. The counting (unless otherwise noted) is by γ-ray spectrometry using 3 × 3 in. solid NaI(Tl) detectors. These sensitivities can be, and are, exceeded by orders of magnitude whenever a particular sample requires it. For example, using a 3 × 3 in. well-type detector in place of the solid detector, the limit of detection for mercury can be lowered from 80 to 3 ng. Using a variety of changes in the analytical procedure, a limit of detection of 5 pg was achieved for one particularly important set of samples.

It is rare that irradiation of a sample results in only the desired nuclear reaction occurring. The usual result is just the opposite. The presence of

TABLE 2

A Periodic Table of the Elements Containing Routinely Achieved Interference-Free Limits of Detection by NAA (μg)[a,b]

1 H NA																	2 He NA
3 Li 0.0008 p	4 Be 15 p											5 B 1.1 p	6 C 0.1 c	7 N 1. c	8 O 1. c	9 F 0.4	10 Ne 2
11 Na 0.004	12 Mg 0.5											13 Al 0.004	14 Si 1. fs	15 P 0.2 b	16 S 4 b fs	17 Cl 0.05	18 Ar 0.002
19 K 0.2	20 Ca 4	21 Sc 0.001	22 Ti 0.1	23 V 0.002	24 Cr 0.3	25 Mn 0.0001	26 Fe 2 fs	27 Co 0.01	28 Ni 0.7	29 Cu 0.002	30 Zn 0.1	31 Ga 0.002	32 Ge 0.1	33 As 0.005	34 Se 0.01	35 Br 0.003	36 Kr 0.01
37 Rb 0.02	38 Sr 0.005	39 Y 0.4	40 Zr 0.8	41 Nb 3	42 Mo 0.1	43 Tc NA	44 Ru 0.04	45 Rh 0.005	46 Pd 0.03	47 Ag 0.004	48 Cd 0.005	49 In 0.00006	50 Sn 0.03	51 Sb 0.007	52 Te 0.03	53 I 0.002	54 Xe 0.1
55 Cs 0.001	56 Ba 0.02	57 La (L) 0.005	72 Hf 0.0006	73 Ta 0.1	74 W 0.004	75 Re 0.0008	76 Os 1	77 Ir 0.0003	78 Pt 0.1	79 Au 0.0005	80 Hg 0.08	81 Tl 1 b	82 Pb 0.5 p	83 Bi 1 b	84 Po NA	85 At NA	86 Rn NA
87 Fr NA	88 Ra NA	89 Ac (A) NA															
		L	58 Ce 0.2	59 Pr 0.03	60 Nd 0.03	61 Pm NA	62 Sm 0.001	63 Eu 0.0001	64 Gd 0.007	65 Tb 0.03	66 Dy 0.00003	67 Ho 0.003	68 Er 0.002	69 Tm 0.2	70 Yb 0.02	71 Lu 0.0003	
		A	90 Th 0.2	91 Pa NA	92 U 0.003	93 Np NA	94 Pu 0.003										

KEY: ATOMIC NUMBER — 33 As — SYMBOL; 0.005 — SENSITIVITY IN MICROGRAMS (INTERFERENCE FREE)

[a] Unless otherwise noted, irradiations are for not longer than 1 hr in a thermal flux of 2×10^{12} n/cm^2 sec.

[b] Key: NA—analysis not performed because of either the natural radioactivity of the element or the lack of a satisfactory nuclear reaction for NAA. p—reactor operated in pulsed mode, b—nuclide counted by beta-ray spectrometry, s—irradiation in a nuclear reactor, with and without cadmium shielding, and f—14-MeV neutron irradiation.

unknown and unwanted radionuclides in a sample intended for β-ray counting is generally fatal to the analysis. When such samples are intended for γ-ray spectrometry, the results may not be so acute. If a photopeak of the desired isotope can be detected in the γ-ray spectrum, the only untoward result will be a lowered analytical precision. If no photopeak is detected, several alternatives exist.

The most direct solution, and usually the only solution if β counting is required, is to perform radiochemical separations. This involves using some scheme, ranging from a classical gravimetric procedure to solvent extractions or ion exchange separations, which will isolate the isotope of interest from interfering activities. A large variety of special procedures make it possible to analyze a sample for an entire group of elements, such as the rare earths, or to look for certain trace elements in a sample whose major constituents are of no particular interest.

Radiochemical separations destroy the sample. If this cannot be tolerated, alternatives can be employed. These may work so well in a particular application that they are preferred to radiochemistry even when the sample is expendable.

"Compton suppression" is a special technique for γ-ray spectrometry which requires that the sample and primary radiation detector be wrapped in a secondary detector. Gamma rays which escape from the primary detector after depositing a fraction of their original energy are responsible for much of the "noise" in a γ-ray spectrum. If the secondary detector is large, the scattered γ ray will probably be captured in its active volume. Output will then be obtained from both detectors simultaneously. By eliminating such events from the γ-ray spectrum, otherwise undetectable photopeaks can be recognized. The elimination process is accomplished by specially designed electronic equipment. See Cooper *et al.* (1969) for a detailed example of the design requirements.

The inverse procedure is "γ–γ coincidence" counting. This is useful when it is desired to detect a single isotope which decays with the simultaneous emission of two γ rays, or for the detection of positron emitting nuclides. The sample is sandwiched between two detectors and only events occurring simultaneously in both are recorded. This procedure is usually refined by electronically analyzing the output of each detector and recording only events in narrow energy intervals centered around the photopeak energies characteristic of the isotope. Johnson (1969) discusses a particularly elegant version of this art.

At times photopeaks may be visualized by "unfolding," "stripping," or "fitting" techniques. These may consist of simply normalizing the sample spectrum to one or more spectra containing unwanted components and manually subtracting the latter from the former. The residual spectrum is then inspected for the desired photopeak. Sophisticated computer programs using least-squares fitting of the sample spectrum with libraries of carefully obtained spectra of pure isotopes are used for the same purpose. Schonfeld *et al.* (1966) have published one computer program using this technique.

2.1.3 *Accuracy*

There is no theoretical reason why the accuracy of NAA results should not equal their precision. For good quality work certain effects should be accounted for in the analytical procedure.

The neutron flux reaching the sample and standard should be identical. The sample and standard may be spun in a neutron beam or rotated about a nuclear reactor core. If both sample and standard are placed in the same irradiation container, "flux monitors" should accompany them to insure that necessary corrections can be made for flux differences at different points in the capsule. A flux monitor is an object which contains one or more elements which will undergo neutron-induced reactions when exposed to neutrons with the same energies as those inducing the reaction desired in the sample. The monitors are counted in some suitable manner and corrections computed for any observed variations in flux.

When γ-ray spectrometry is to be employed, γ-ray attenuation within the sample and standard should either be identical or correction factors should be determined. It is obviously wrong to use an aqueous standard with a sample whose matrix is largely lead if the γ ray to be detected is of low energy and the sample is more than a few millimeters thick. Ingenuity is sometimes required in obtaining appropriate standards or in making appropriate corrections. These effects can be neglected if radiochemistry is employed since both sample and standard can be counted in identical states.

Neutron absorption or thermalization effects should be considered. For example, the chlorine in 1 g of table salt will absorb a sufficient fraction of incident neutrons that the thermal neutron flux inside the sample is significantly depressed. The standard should be either prepared in a similar matrix or experiments be performed to determine an appropriate correction. Similarly, large samples of hydrogenous material may thermalize neutrons to such an extent that the thermal neutron flux inside the sample is significantly enhanced.

In specific instances errors can occur owing to a change in chemical state which occurs in the element to be determined during irradiation. Atoms which have participated in nuclear reactions recoil with high kinetic energies and exist in very reactive chemical states. For example, if a liquid sample containing chloroform is irradiated, some of the chlorine which undergoes the $^{37}Cl(n, \gamma)^{38}Cl$ reaction may be present at the end of the irradiation as HCl or as Cl_2. These species may volatilize from the sample or be adsorbed to the walls of the irradiation container and be lost when the sample is transferred to an unirradiated container for counting. Losses can be acute in gaseous samples. For example if a carbon dioxide sample containing traces of arsine is irradiated it can be expected that essentially all of the ^{76}As produced will be deposited to the walls of the irradiation container.

Systematic errors arising from sample contamination are largely unknown in NAA. If the sample can be kept uncontaminated until the end of irradiation, only the addition of radioactive contaminants can affect the accuracy of the final result. Even in laboratories engaged heavily in handling radioactive materials, it is relatively simple to insure that such events do not occur.

2.1.4 *Precision*

When a single channel analyzer, such as a β detector plus scaler is used to count a radioactive sample, the standard deviation of a measurement consisting of n_r counts accumulated in time t is

$$\sigma = n_r^{1/2} \tag{13}$$

If the background for the same counter is determined by counting a nonradioactive sample for time T, accumulating n_b counts, the percent relative standard deviation of the net counts from the sample is

$$\sigma_r = 10^2[n_r + (t^2/T^2)n_b]^{1/2}/[n_r - (t/T)n_b] \tag{14}$$

If a standard is counted and the precision of the count calculated by (14) to be σ_s, then σ, relative standard deviation of the entire analytical procedure is given by

$$\sigma = (\sigma_r^2 + \sigma_s^2)^{1/2} \tag{15}$$

In γ-ray spectrometry, many different mathematical techniques are used to quantify the results. One popular and straightforward approach is simply to integrate the counts in the peak and subtract a background determined from the counts on each side of the peak. In this case the precision of the analysis is computed as in (14) except that the number of channels used in integrating the peak c replaces the time that the sample was counted, t. Similarly, the number of channels used to compute the background C replaces the time the background was counted, T. The resulting equation is

$$\sigma_r = 10^2[n_r + (c^2/C^2)n_b]^{1/2}/[n_r - (c/C)n_b] \tag{16}$$

Other techniques for quantifying results obtained from γ-ray spectrometry lead to different expressions for the precision of the analysis.

In favorable cases, where the indicator isotope is present in abundance and is of relatively long half-life, and where no interferences exist, $n_r \geq (c^2/C^2)n_b$ and $n_r \gg (c/C)n_b$. Then from (16) we obtain

$$\sigma_r \sim 10^2(2n_r)^{1/2}/n_r \sim 140n_r^{-1/2} \tag{17}$$

This implies that if 10^4 counts can be accumulated in channels centered around the photopeak, σ_r should be less than 2%. Presently available equipment permits 10^4 counts to be accumulated in 1 min and 10^4 counts in the photopeak area can be obtained in a few minutes of counting.

Most multichannel pulse-height analyzers now in use contain memories capable of storing $10^5 - 1$ counts per channel so that values of n_r exceeding 10^6 are quite feasible. Thus precisions exceeding 0.2% are attainable.

2.1.5 *Size and Kind of Specimen*

NAA has been performed on samples barely visible to the naked eye and on samples several square feet in area. Every imaginable type of material from floor wax to lunar rocks has been analyzed. Solid, liquid, and gaseous samples are suitable.

When very small samples are analyzed, the major problem is to optimize irradiation and counting conditions so that adequate sensitivities are obtained. With very large samples it may be difficult to insure that the neutron flux over the entire sample is uniform. It may be possible to move the sample relative to the neutron source. For example, bulk samples can be pumped or passed on a conveyor through irradiation and counting positions. A large solid sample can be analyzed by moving it over a beam port emerging from a reactor core in some systematic fashion or by irradiating small areas with a portable neutron source and averaging the results. Counting very large samples may also require movement of the sample relative to the detector. Considerable ingenuity may be required to prepare adequate standards when large samples are to be analyzed. Both the neutronic and γ-ray effects discussed in Section 3 can become acute. The problem is exacerbated when the sample is nonhomogeneous, as for example in the analysis for calcium in an animal's body.

2.1.6 *NDT or DT*

As was pointed out in Section 1, many techniques have been developed for nondestructive, or instrumental, NAA. In certain instances no successful instrumental approach is available. Other than turning to some other method of analysis, the only recourse is to radiochemically isolate the activity of interest. Because of the small sample size required for NAA, such a procedure may hardly qualify as a destructive analysis. Seldom is more than 1 g of material required for analysis and in some cases much less suffices. Paint samples weighing less than 10 μg were removed from a potentially valuable painting and successfully analyzed for certain major constituents by NAA.

When radiochemical separations are performed, it is common practice to add large excesses of stable isotopes of the element to be determined to the irradiated sample. This serves the dual purpose of transforming the analysis and of providing a method for detecting the loss of material during the analysis. For example, if selenium is to be analyzed in 1 g of tissue, the sample is irradiated by thermal neutrons to induce the production of ^{75}Se, and 10 mg of stable selenium are added to the sample which is digested in hot acids. After some chemical procedures metallic selenium is isolated which contains both radioactive ^{75}Se and the stable isotopes. The radioactive ^{75}Se is detected by γ-ray spectrometry. The loss of ^{75}Se through the chemical procedures is inferred by weighing, or otherwise analyzing, the counted sample for its selenium content.

Changes are induced in samples by exposing them to neutrons. The number of isotopic transmutations, while measurable, is hardly significant. For example, if 1 cm^3 of a pure isotope, containing 10^{22} atoms, is irradiated for 10^4 sec in a neutron flux of 10^{13} neutrons/cm^2 sec, and the isotope has the large capture cross section of 10^{-22} cm^2, the total number of transmutations which take place is 10^{17}. Thus 0.001% of the sample is transmuted is this extreme case. In the majority of cases residual radioactivity in the sample is not a significant problem. When the analysis is completed by radiochemical procedures the radioactive wastes are disposed of in a prescribed manner. In nondestructive analyses the amount of activity induced in the sample need only be that required to achieve a desired response in the radiation detector. Retention of the sample for some time will allow the induced activity to decrease to nondetectable levels. In a few cases it may be necessary to produce significant amounts of long-lived isotopes. Appropriate state and federal regulations delineate the responsibility of the analyst in such instances.

Some changes other than those caused by neutron capture reactions may occur. Transparent samples such as gems or glasses may be colored by radiation-caused electron dislocations. These colors are readily removed by heating the material to temperatures below the softening or melting point of the material.

Biological and organic samples reflect changes in chemical composition by changes in morphology and color after long irradiations at high fluxes.

Similarly organic polymers tend to harden owing to cross-linking reactions induced by ionizing radiation. Organic monomers such as styrene or unsaturated oils polymerize by similar mechanisms. Lattice defects caused by various forms of radiation occur in crystalline materials.

2.1.7 *Time Required for Analysis*

This factor is strongly influenced by the particular isotope to be determined and by the sample matrix. Some analyses are completed in less than a minute, including both irradiation and counting. Some analyses may require weeks to complete. If radiochemistry is not required, the analyst's time required is not much different in either case, making NAA an economical technique in terms of man-hours per sample.

2.2 Examples of Applications of NAA

A recent bibliography compiled by Lutz *et al.* (1971) lists 9480 publications concerned with activation analysis. While some of the articles cited deal with photonuclear or charged particle activation analysis, the majority concern NAA. It is impossible to adequately summarize so large a body of work in a few pages. A relatively few applications are selected as illustrative of a particular feature of the method or as of wide interest.

2.2.1 *The Determination of Oxygen in Beryllium*

Dissolved oxygen or oxide inclusions have adverse effects on the physical properties of beryllium and many other metals. The need for a rapid, accurate method of analysis is met by NAA. The method described here is summarized from McCrary *et al.* (1961) and is applicable to other metals, alloys, and many other kinds of samples with only minor modifications. A more recent system, suitable for very accurate determinations, is described by Lundgren and Nargowalla (1968).

The reaction used is the $^{16}O(n, p)^{16}N$ fast-neutron reaction, which has a threshold of 10.0 MeV. Cockcroft-Walton neutron generators are very suitable neutron sources. Deuterium ions are accelerated and impinged on a tritium-loaded target, producing 14-MeV neutrons by the $^{3}H(d, n)^{4}He$ reaction. Fast neutron fluxes of about 2×10^8 neutrons/cm^2 sec are delivered to the sample, giving limits of detection of about 15 ppm for a 20-g sample. Time-dependent flux changes occur and the flux is monitored via a plastic scintillator placed behind the sample and coupled to a photomultiplier tube. The photomultiplier tube output is amplified and fed to a discriminator adjusted so that only pulses originating from a proton recoil (caused by scattering after a neutron-proton collision) of at least 7-MeV energy are recorded in a scaler.

The product isotope (^{16}N) has a half-life of 7 sec and decays with emission of 6 and 7 MeV γ rays. The short half-life is a strong inducement to automate the irradiation and counting operations. A pneumatic transfer system operated by timing circuits is used. The only manual operations are the loading and unloading of the samples from the system.

In the system described a 3 × 3 in. NaI(Tl) detector was used. Sensitivity is enhanced if dual 5 × 5 in. detectors are opposed and the sample counted between the detectors. Also, a 256-channel pulse-height analyzer is described which could be replaced by a discriminator-scaler.

Samples, standards, and blanks are irradiated and counted in low-oxygen plastic containers. The observed oxygen activity in each is normalized to the flux monitor counts, blank corrections are applied, and the oxygen content of the sample calculated in the usual manner by comparison with the standard.

Possible interferences are from boron and fluorine in the sample. Boron interferences arise from the $^{11}B(n, p)^{11}Be$ reaction. The product nuclide has a half-life of about 14 sec and decays with the emission of 5.85, 6.79, and 7.99 MeV γ rays. Flourine produces ^{16}N by the $^{19}F(n, \alpha)^{16}N$ reaction. If the levels of these elements in the sample can be determined by an alternate analytical procedure, corrections can be applied. The production of ^{16}N by the $^{15}N(n, \gamma)^{16}N$ reactions is insignificant in 14-MeV neutron irradiations.

Portable systems similar to the one described have been constructed for use in geological exploration and in industrial process control. Peteu (1969) has summarized this technology.

2.2.2 *Multielement Scanning of Crude and Residual Oils*

Multielement scanning is one of the most generally useful techniques in NAA. Samples of almost every conceivable description have been analyzed and useful information obtained. The essentials are quite simple; the sample is irradiated one or more times and counted one or more times after each irradiation. Photopeaks are associated with specific isotopes and quantitative analyses made. Because of the high thermal fluxes available, nuclear reactors are usually employed for irradiations.

The goal of the procedure is to detect and quantify as many elements as possible in the sample. Where an element of interest is not detected, "upper limits"—the maximum amount of the element which could be present and escape detection—can be readily computed.

The procedure described here was used in a study directed toward identifying the source of oil slicks by Lukens *et al.* (1971), but it differs only in detail from procedures applied to other samples.

About 1 g of each sample is irradiated for about 1 min in a thermal neutron flux of 2.8×10^{12} neutrons/cm^2 sec. Counting commences 1 min after irradiation and continues for 1 min (live time). A 3 × 3 in. NaI(Tl) detector is coupled to a 400-channel pulse-height analyzer adjusted to cover the energy range 0–3 MeV. A 0.5-in. plastic disk is interposed between the sample and detector to filter β rays from the detector. Characteristic γ rays from activities with short half-lives are detected. A vanadium standard is irradiated and counted with each set of samples and serves as a flux monitor.

About 5 g of each sample are then irradiated for 30 min in a thermal flux of 1.8×10^{12} neutrons/cm^2 sec. Counting commences 1 hr and 1 day after irradiation. Counts are of 15 and 90 min duration (live time), respectively. A 36-cm^3 Ge(Li) detector coupled to a 4096-channel pulse-height analyzer is employed and the region 0–2 MeV scanned. An arsenic standard serves as a flux monitor. Activities with half-lives ranging from several minutes to several years are observed.

In 374 samples analyzed, 35 elements were detected at least once and 16 of these were detected with frequencies ranging from 100 to 18% (these were V, Br, S, Mn, Ni, I, Al, Ga, As, Zn, Ba, Co, Na, Cl, In, and Dy).

The analytical procedure and the data reduction were largely automated in this work. An automatic sample changer was used in conjunction with the 4096-channel analyzer. Analyzer output was transferred to either punched cards or magnetic tape. A high-speed digital computer analyzed each spectrum, reporting the energy of peaks and identifying probable isotopes associated with each peak. The computer was then used to determine the amount of each element detected and to derive upper limits for elements not detected.

2.2.3 *Some Examples of Single Element Analysis by Instrumental NAA*

Some analyses can be performed very quickly and economically by NAA. These lend themselves to routine applications. Three examples are given.

a. The Detection of Fluorine in Fabrics. Fluorine-containing polymers are widely used to impart water- and stain-resistant properties to fabrics. Conventional chemical procedures for the detection of this element are tedious and subject to systematic errors due to the difficulty in decomposing the polymer and avoiding losses of volatile decomposition products.

The reaction $^{19}F(n, \gamma)^{20}F$ occurs in good yield in nuclear reactor irradiations. The product isotope has a half-life of 11.4 sec and decays with the emission of a 1.64-MeV γ ray. Other sources of the product isotope, the $^{23}Na(n, \alpha)^{20}F$ and $^{20}Ne(n, p)^{20}F$ reactions, can be neglected in these samples, since experiments confirm that these reactions have relatively poor yields and the parent isotopes are in relatively low concentration.

About 3 g of fabric are irradiated in thermal neutron fluxes of about 10^{11} neutrons/cm^2 sec for about 15 sec. The samples are counted for 12 sec, beginning 12 sec after irradiation, using 3 × 3 in. NaI(Tl) detectors and 400-channel pulse-height analyzers. Aluminum is usually present as a trace element in fabric and yields 2.3-min ^{28}Al, which decays with the emission of a 1.78-MeV γ ray.

To diminish the interference from ^{28}Al activity, the spectrum accumulated during the initial count is stripped. This is accomplished by switching the analyzer from the "add" to "subtract" mode after the initial count is completed. Then, exactly 1 min after the first count was begun, with the accumulated spectrum still in the analyzer memory and the sample in place, a second count is begun. Activities with half-lives longer than ^{20}F, especially ^{28}Al, are still present in the sample in good abundance and subtract themselves from the analyzer memory. The 1.64-MeV photopeak remains nearly undiminished and relatively free of interference. The standard is counted exactly as the samples. Data reduction is assisted by the use of a high-speed digital computer.

b. Vanadium in Crude Oil. Vanadium is present in crude oil in widely varying concentration. It tends to poison cracking catalysts so that its determination is important in refinery operation.

The 1.43-MeV γ ray of 3.75-min ^{52}V is the most prominent photopeak observed in a spectrum obtained shortly after the irradiation of a crude oil sample in a thermal neutron flux. The determination of vanadium is very simple by the $^{51}V(n, \gamma)^{52}V$ reaction and the poor yields of competing reactions [$^{52}Cr(n, p)^{52}V$ and $^{55}Mn(n, \alpha)^{52}V$] allow them to be ignored as possible sources of error.

The recent availability of ^{252}Cf neutron sources has led to the construction of an on-line NAA system capable of continuously monitoring the concentration of vanadium in refinery feedstock.

c. High-Precision Analysis of Uranium Ores. Uranium exploration and mining is a large and growing industry. A need exists for rapid, accurate determinations of the element. Conventional analytical methods are relatively tedious.

The basis for the analytical procedure is the delayed neutron phenomenon discussed in the section on selectivity (2.1.1). A nuclear reactor serves as the neutron source and the detector consists of a $^{10}BF_3$ neutron detector surrounded by about 1 ft^3 of polyethylene. The polyethylene moderates the neutrons emanating from the sample and these thermal neutrons induce the

$^{10}B(n, \alpha)^7Li$ reaction which is the basis for detector operation. These detectors are essentiallly insensitive to γ rays and are shielded from β^- rays and externally produced α rays by the experimental design. The irradiated sample terminates in the center of the polyethylene block with the detector offset about 2 in. The long axes of the detector and sample are parallel, with the sample positioned at the detector midline.

With 15-sec irradiations in a thermal flux of 2×10^{12} neutrons/cm^2 sec, 10 ppm of uranium in a 0.3-g sample can be detected. Exactly 3 sec are allowed for the sample to return from the reactor core and come to rest at the counting position. Detector counts are accumulated for 15 sec in a conventional discriminator-scaler. The entire irradiation and counting sequence is automated. Ores containing 0.1% uranium can be analyzed with precisions (1σ) of $\pm 1.5\%$ of the value and those containing 1.0% uranium with precisions of $\pm 0.5\%$. The system has been used to analyze reference ore samples ranging from 0.17 to 0.38% in uranium. By repetitively analyzing the samples, results differing by less than $\pm 0.5\%$ from the accepted values were obtained.

With some refinements, this technique has been adapted to the analysis of reactor fuels with overall accuracies better than $\pm 0.5\%$ (1σ) consistently achieved.

2.2.4 *The Determination of Rare-Earth Element Group Separations after Neutron Activation*

The relative concentrations of rare-earth elements in minerals are of scientific interest to geochemists. In addition, the increasing economic importance of rare earths has created a need for simple, accurate analyses. The procedure summarized from Graber *et al.* (1970) describes a radiochemical procedure for isolating these elements from rocks and their determination. If high quality ores or purified rare earths are to be analyzed a nondestructive analysis can be performed.

In a thermal flux of 2×10^{12} neutrons/cm^2 sec 1-g portions and standards of each rare-earth element are irradiated for 30 min. The irradiated samples are placed in crucibles, 10 mg of stable lanthanum are added, then several grams of sodium carbonate. The samples are fused in a furnace and the resulting mass is dissolved in dilute nitric acid. Precipitations of rare-earth flourides and hydroxides are made which isolate the rare-earth elements. The chemical separation requires about 2 hr. The purified rare earths are dissolved in dilute nitric acid and are counted using a 40-cm^3 Ge–Li detector and 4096-channel analyzer. Repetitive counts are made (as in multielement scans) so that longer lived nuclides can be detected in the sample after the shorter lived activities decay. The first count is made on the day of irradiation and 2.3-hr ^{165}Dy and 7.5-hr ^{171}Er detected. From the count made the next day 9.3-hr $^{152}Eu^{m}$, 18-hr ^{159}Gd, 40-hr ^{140}La, 19-hr ^{142}Pr, and 27-hr ^{166}Ho are determined. Two days later 33-hr ^{143}Ce, 6.7-day ^{177}Lu, 101-hr ^{175}Yb, and 47-hr ^{153}Sm photopeaks are visible. Eight days after irradiation 11-day ^{117}Nd and 72-day ^{160}Tb activities reveal themselves. Detection of the last rare-earth isotope, 134-day ^{170}Tm, requires that the sample be counted 26 days postirradiation.

Standards are counted at appropriate times and the usual techniques used to calculate the amounts of rare earth in each sample. Blanks are not required.

After the last count, the "radiochemical yield" must be determined. This term refers to the fact that some rare earth is lost during the radiochemical procedure and therefore the samples counted represent only a fraction of the original amount present. Since rare-earth elements are chemically very similar, it is only necessary to determine how much of the 10 mg of lanthanum added to each sample remains and divide that fraction into all of the values obtained.

It is convenient to determine the lanthanum by NAA. A 15-sec irradiation in a thermal flux of 2×10^{12} neutrons/cm^2 sec provides a plentiful supply of 1.59-MeV γ rays from the decay of ^{140}La. The ^{140}La activity in each sample is compared with a standard consisting of 10.0 mg of La, irradiated with the samples.

It should be emphasized that while the analytical procedure described here may take a month to complete, all of the manipulations except the radiochemistry can either be automated or left to technicians. Overall costs are competitive with more conventional methods of analysis.

3 Data Form and Interpretation

Essentially all routine neutron activation analysis done results in the data being presented in the form of a γ-ray spectrum. The spectrum is merely a plot of γ-ray energy on the abscissa versus counts per channel on the ordinate. The spectra are obtained with a thallium-activated NaI scintillation detector, NaI(Tl), coupled to a multichannel γ-ray spectrometer (typically 400 channels) or with a high-resolution lithium-drifted germanium detector, Ge(Li), used in conjunction with a 4096 channel γ-ray spectrometer. The germanium detector offers increased resolution over the NaI detector at the expense of counting efficiency. For the purposes of data interpretation, however, the two spectra can be treated the same. The spectrum is displayed on an oscilloscope on the spectrometer. For convenience of interpretation the data can be obtained in digital form on printed or punched paper tape, punched cards, or magnetic tape. The spectrometers can also be interfaced to small computers for rapid data reductions (see Chapter 20).

If the counting equipment is properly calibrated, the response is linear over the range of γ-ray energies of interest, usually from 0 to 3 MeV. This means the calibration curve is a straight line going through the origin. If this is the case, each channel in the spectrum can be assigned an energy increment which is constant, greatly simplifying the identification of the energy of the various peaks in the spectrum. The first step in the identification of the peaks is to determine the calibration of the spectrometer. This is done by counting standard isotopes of known γ-ray energies and determining the energy increment per channel by dividing the energy of the γ ray expressed in keV by the channel number in which the peak is found. The result, called the gain and expressed in keV per channel, should be the same for at least two γ rays of different energies which cover approximately the energy range of interest. If this is the case, the energy of any unknown peak can be determined simply by multiplying the gain by the channel number. If the gain is found to be different for the different energies of the standard isotope, then the calibration curve does not go through the origin (i.e., zero energy does not correspond to channel zero) and the spectrometer should be recalibrated. If this is not possible, the energy of an unknown peak can still be determined by plotting the calibration curve and reaching the energy directly from the curve. Since several of certain iotopes may have γ rays of approximately the same energy, energy identification is sometimes only the first step in identifying the unknown isotope. To positively identify the isotope, its half-life may also be determined. This is done by counting the sample at least three times with appropriate delays, these being determined by the known half-lives of possible candidates with the appropriate energies. The logarithms of the counts per minute obtained are then plotted versus time to obtain the decay curve. If this curve is a straight line, then there are no interfering activities contributing to the peak of interest and the half-life can be determined from the plot simply by noting the time required for the count rate to be cut in half. With the peak energy and half-life data, identification is accomplished by consulting one of various compilations of γ energies and half-lives of radioactive isotopes. The use of the high-resolution Ge(Li) detector allows the peak energy to be determined to at least three significant figures. In most cases, this accurate determination is sufficient to identify the isotope without making a half-life determination. A listing of the more important γ rays encountered in NAA is given in Table 3.

Most radioactive isotopes emit several γ rays of various energies and intensities, a feature which greatly aids in the interpretation of the spectra.

TABLE 3

THE MORE IMPORTANT γ RAYS ENCOUNTERED IN NAA SUBSEQUENT TO ACTIVATION WITH A NUCLEAR REACTOR

Energy (keV)	Isotope	Half-life[a]	Energy (keV)	Isotope	Half-life[a]	Energy (keV)	Isotope	Half-life[a]
50	^{169}Yb	32d	129	^{129}Os	16d	208	^{177}Lu	6.8d
51	^{104}Rh^{m}	4.4m		^{195}Pt^{m}	3.5d		^{199}Au	3.05d
53	^{183}W^{m}	5.5s	130	^{197}Au^{m}	7.2s	210	^{77}Ge	11.3h
57	^{143}Ce	33.4h	133	^{181}Hf	44.6d		^{149}Nd	1.8h
58	^{159}Gd	18.5h	134	^{197}Hg^{m}	24h	214	^{178}Hf^{m}	4.8s
59	^{60}Co^{m}	10m	136	^{75}Se	120d	215	^{97}Ru	2.8d
63	^{169}Yb	32d	137	^{186}Re	88.9h		^{180}Hf^{m}	5.5h
70	^{153}Sm	47h	139	^{75}Ge^{m}	49s	217	^{179}Hf^{m}	19s
72	^{187}W	24h		^{193}Os	30.6h	228	^{177}Yb^{m}	6.5s
75	^{239}U	23.5m	140	^{99}Tc^{m}	6h		^{239}Np	2.35d
77	^{197}Pt	18h	141	^{101}Mo	14.6m	230	*^{132}Te	77.7h
	^{197}Hg	65h	142	^{46}Sc^{m}	19.5s	231	^{85}Sr^{m}	70m
80	^{131}I	8.05d	145	^{141}Ce	32.5d	246	^{155}Sm	24m
81	^{166}Ho	27h	147	^{182}Ta^{m}	16.5m	247	^{111}Cd^{m}	49m
87	^{233}Th	23.3m	150	^{85}Kr^{m}	4.4h	250	^{135}Xe	9.13h
88	^{109}Pd	13.6h		^{111}Cd^{m}	49m	263	^{77}Ge	11.3h
	^{176}Lu^{m}	3.7h		^{131}Te	25m	265	^{75}Ge	82m
91	^{147}Nd	11.1d	155	^{188}Re	17h		^{75}Se	120d
92	^{188}Re^{m}	18.7m	158	^{199}Au	3.15d	270	^{149}Nd	1.8h
95	^{165}Dy	2.3h	160	^{123}Sn^{m}	41m	273	^{117}Cd^{m}	3.4h
99	^{183}W^{m}	5.5s	161	^{77}Se^{m}	17.5s	278	^{239}Np	2.35d
	^{195}Pt^{m}	3.5d	166	^{139}Ba	84m	279	^{197}Au^{m}	7.2s
102	^{161}Gd	3.7m	170	*^{151}Pm[b]	27.8h		^{197}Hg^{m}	24h
103	^{81}Se^{m}	57m	172	^{182}Ta^{m}	16.5m		^{203}Hg	46.9d
	^{153}Sm	47h	174	^{151}Nd	12m	280	^{75}Se	120d
104	^{155}Sm	24m	181	^{99}Mo	67h		^{193}Os	30.6h
	^{177}Yb^{m}	6.5s	184	^{182}Ta^{m}	16.5m	283	^{175}Yb	4.2d
106	^{188}Re^{m}	18.7m	188	^{109}Pd^{m}	4.8m	284	^{131}I	8.05d
	^{239}Np	2.35d	190	^{81}Kr^{m}	13s	286	*^{149}Pm	53.1h
108	^{165}Dy^{m}	1.3m	191	^{101}Mo	14.6m	293	^{143}Ce	33.4h
	^{183}W^{m}	5.5s		^{197}Hg	65h	296	^{171}Er	7.5h
112	^{171}Er	7.5h		^{197}Pt	18h	299	^{160}Tb	73d
113	^{177}Lu	6.8d	192	^{114}In^{m}	50d	305	^{85}Kr^{m}	4.4h
114	^{149}Nd	1.8h	197	^{19}O	29s		^{140}Ba	12.8d
118	^{151}Nd	12m		^{199}Pt	31m	306	^{105}Rh	35.9h
122	^{152}Eu^{m}	9.3h	198	^{169}Yb	32d	307	^{101}Te	14m
	^{152}Eu	12.3y	202	^{90}Y^{m}	3.2h	308	^{171}Er	7.5h
128	^{134}Cs^{m}	2.9h	208	^{167}Er^{m}	2.5s		^{192}Ir	74.4d

[a] Abbreviations used here are as follows: s—second; m—minute, h—hour; d—day; y—year.

[b] Isotopes usually derived only from fission are denoted by an asterisk.

Table 3 continued

Energy (keV)	Isotope	Half-life[a]	Energy (keV)	Isotope	Half-life[a]	Energy (keV)	Isotope	Half-life[a]
312	^{233}Pa	27.0d	482	^{181}Hf	44.6d	632	^{108}Ag	2.3m
315	^{161}Gd	3.7m	487	^{140}La	40.2h	640	^{194}Ir	19h
316	^{192}Ir	74.4d	490	^{115}Cd	53h	644	^{124}Sb^{m}	1.3m
319	^{105}Rh	35.9h	497	^{103}Ru	40d	658	^{110}Ag^{m}	270d
320	^{51}Ti	5.8m	502	^{190}Os^{m}	9.5m		^{110}Ag	24s
	^{51}Cr	27.8d	505	^{124}Sb^{m}	1.3m	660	^{76}As	26.5h
325	^{125}Sn^{m}	9.5m	511	^{13}N	10m	662	^{137}Ba^{m}	2.6m
326	^{178}Hf^{m}	4.8s		^{15}O	124s		*^{137}Cs	30y
328	^{194}Ir	19h		^{18}F	1.87h	665	^{97}Nb	74m
329	^{140}La	40.2h		^{30}P	2.5m	666	^{80}As	15.3s
333	^{180}Hf^{m}	5.5h		^{64}Cu	12.5h	668	^{143}Ce	33.4h
340	*^{151}Pm	27.8h		^{65}Zn	245d	670	^{105}Ru	4.5h
344	^{152}Eu	12.7y		^{80}Br	17.6m		*^{132}I	2.26h
346	^{181}Hf	44.6d		^{101}Mo	14.6m	676	^{198}Au	2.7d
361	^{161}Gd	3.7m		^{108}Ag	2.3m	686	^{122}Sb	2.8d
	^{165}Dy	2.3h		^{140}Pr	3.4m		^{187}W	24h
	^{190}Os^{m}	9.5m	512	^{106}Rh	2.2h	690	^{129}Te^{m}	34.1d
363	^{159}Gd	18.5h	514	^{165}Dy^{m}	1.3m	720	^{124}Sb	60d
364	^{131}I	8.05d	527	^{135}Xe^{m}	15.6m	724	^{95}Zr	65d
368	^{65}Ni	2.56h	528	^{128}I	25m		^{114}In^{m}	50d
375	^{204}Pb^{m}	66.9m	530	^{115}Cd	53h	726	^{105}Ru	4.5h
388	^{87}Sf^{m}	2.8h		*^{133}I	20.3h	740	^{99}Mo	67h
396	^{175}Yb	4.2d	533	^{147}Nd	11.1d	747	^{97}Zr	17h
403	^{87}Kr	78m	537	*^{140}Ba	12.8d		^{97}Nb^{m}	1m
412	^{198}Au	2.7d	540	^{199}Pt	31m	756	^{95}Zr	65d
417	^{77}Ge	11.3h	545	^{101}Tc	14m	765	^{95}Nb	35d
	^{116}In^{m}	54m	550	^{82}Br	35.8h	773	*^{132}I	2.26h
427	^{178}Hf^{m}	4.8s	559	^{76}As	26.5h	777	^{82}Br	35.8h
430	^{147}Nd	11.1d	560	^{86}Rb^{m}	1m	779	^{152}Eu	12.3y
434	^{108}Ag	2.3m		^{104}Rh	425	780	^{99}Mo	67h
439	^{23}Ne	38s	564	^{122}Sb	2.8d		^{131}Te^{m}	1.2d
	^{69}Zn^{m}	14h	570	^{134}Cs	2.07y	796	^{134}Cs	2.07y
441	^{128}I	25m	590	^{101}Mo	14.6m	835	^{54}Mn	291d
444	^{180}Hf^{m}	15.5h	603	^{124}Sb^{m}	1.3m		^{72}Ga	14.1h
453	^{131}Te	25m		^{124}Sb	60d	840	^{27}Mg	9.5m
	^{233}Th	23.3m	605	^{134}Cs	2.07y	842	^{152}Eu^{m}	9.3h
455	^{129}Te	74m	610	^{103}Ru	40d	847	^{56}Mn	2.58h
460	^{193}Os	30.6h		*^{134}I	52m	850	^{87}Kr	78m
468	^{192}Ir	74.4d	616	^{190}Os^{m}	9.5m		*^{134}I	52m
475	^{105}Ru	4.5h	618	^{80}Br	17.6m	871	^{94}Nb^{m}	6.6m
	^{199}Pt	31m	620	^{82}Br	35.8h	879	^{160}Tb	73d
479	^{187}W	24h	622	^{106}Rh	30s	885	^{110}Ag^{m}	270d
482	^{90}Y^{m}	3.2h	630	^{72}Ga	14.1b	889	^{46}Sc	83.5d

Table 3 continued

Energy (keV)	Isotope	Half-life[a]	Energy (keV)	Isotope	Half-life[a]	Energy (keV)	Isotope	Half-life[a]
890	*^{134}I	52m	1140	*^{135}I	6.68h	1596	^{140}La	40.2h
894	^{72}Ga	14.1h	1170	^{60}Co	5.27y	1600	^{38}Cl	37.3m
898	^{88}Rb	17.8m	1220	^{76}As	26.5h	1630	^{20}F	11s
900	^{204}Pb^{m}	66.9m	1222	^{182}Ta	115d	1640	^{80}As	15.3s
910	^{89}Y^{m}	16.1s	1260	^{31}Si	2.62h	1692	^{124}Sb	60d
955	*^{132}I	2.26h	1280	^{29}Al	6.6m	1780	^{28}Al	2.3m
963	^{152}Eu^{m}	9.3h		*^{135}I	6.68h	1810	^{56}Mn	2.58h
966	^{160}Tb	73d	1290	^{41}Ar	1.82h	1863	^{88}Rb	17.8m
1010	^{27}Mg	9.5m		^{59}Fe	45.1d	2110	^{56}Mn	2.58h
1020	^{101}Mo	14.6m	1293	^{116}In^{m}	54m	2130	^{34}P	12.4s
1039	^{66}Cu	5.1m	1330	^{60}Co	5.27y	2170	^{38}Cl	37.3m
1078	^{86}Rb	19.5d	1370	^{19}O	29s	2570	^{87}Kr	78m
1090	^{116}In^{m}	54m		^{24}Na	15h	2740	^{24}Na	15h
1095	^{59}Fe	45.1d	1380	^{166}Ho	27h	3090	^{37}S	5m
1110	^{65}Ni	2.56h	1430	^{52}V	3.8m	3100	^{49}Ca	8.8m
1115	^{65}Zn	245d	1480	^{65}Ni	2.56h	4100	^{49}Ca	8.8m
1120	^{46}Sc	83.5d	1520	^{42}K	12.5h	6130	^{16}N	7.14s
1122	^{182}Ta	115d	1570	^{142}Pr	19.1h	7110	^{16}N	7.14s

Identification can sometimes be simplified by looking for other peaks associated with possible isotopes giving rise to the peak of interest in the spectrum. As an example, in the absence of a half-life determination, ^{27}Mg (major γ ray of 844.0 keV) can be distinguished from ^{56}Mn (major γ ray of 846.9 keV) by the presence of a 1014.1 keV peak in the spectra of the former and 1810.7 and 2112.8 keV peaks in the spectra of the latter. Characteristics of the decay scheme of isotopes can also aid in their identification. A classic example is that of ^{59}Fe (γ rays of 1098.6 and 1291.5 keV, half-life of 45 days) and ^{60}Co (γ rays of 1173.1 and 1332.4 keV, half-life of 5.24 years). When counted on a NaI(Tl) detector, the spectra of these isotopes look almost identical and their half-lives are too long to conveniently make a half-life determination. However, the ^{60}Co γ rays are emitted in coincidence and those of ^{59}Fe are not. As a result, a peak, especially prominent when counted in a well-type NaI(Tl) crystal, appears at 2505.5 keV in the ^{60}Co spectrum which is absent in the ^{59}Fe spectrum. This peak, called a sum peak, does not represent a true γ ray of ^{60}Co, but rather is the sum of the energies of the two γ rays, arising from the coincidental loss of the total energy of both ^{60}Co γ rays in the detector. Other phenomena which should be watched for when interpreting γ-ray spectra are escape peaks and double escape peaks, discussed in the theoretical section.

In addition to being a qualitative method, activation analysis is also quantitative. The area under the γ-ray photopeak is directly proportional to the specific activity of the induced radiation and hence to the amount of the element in the sample. By irradiating and counting standards of known composition for each element of interest under identical conditions to those of the sample, the amounts of the elements in the sample can be determined. The peak areas must be expressed as a counting rate, usually counts per minute, obtained by dividing the area by the length of time the sample or standard was counted, and they must be corrected for radioactive decay to the same arbitrary time. The peaks must also be free of interfering peaks of about the same energy.

Several methods have been suggested to determine the area of the peak. Perhaps the simplest is to determine from the digital information the channel on both sides of the peak which has the minimum number of counts and to sum the counts in the channels between these two minima. This will give the gross counts in the peak. In most cases, however, the peak will be superimposed on a rather high background and this base area must be determined and subtracted from the gross area to obtain the net peak area, the entity used in the quantitative measurements. The base area is trapezoidal and can be determined by averaging the number of counts in the two minima channels and multiplying by the number of channels in the peak. A better measurement of the base can usually be obtained by averaging the counts in several channels immediately adjacent to the peak rather than using just one on each side.

The use of high-speed computers greatly increases the speed with which routine data interpretation can be done as well as allowing solutions to complex data problems. Programs are available to produce plots of the spectra using a computer in conjunction with a plotter or other data graphic device. There are also programs which will scan a spectrum, locate the peaks, determine their energy, and assign possible identifications based on a table of isotope energies supplied in the program. After peak identification is made, the concentration of the elements of interest in the sample as well as the more tedious calculation of the standard deviation in the result is quickly computed. The computer is supplied the spectra of the samples and standards and other necessary information such as irradiation times, counting times, sample weights, half-lives, etc. All integrations, decay corrections, and statistical calculation are done by the computer, and the results are printed as parts per million.

All data interpretations described thus far are applicable to routine calculations on interference-free samples. With the advent of the Ge(Li) detector, almost all calculations are of this type. There are cases, however, where two or more peaks are not resolved or where an apparently single peak is actually attributable to more than one isotope of different half-lives, resulting in a complex decay curve. The computer has become a valuable tool to resolve these complex data-handling problems.

References

Adams, F., and Dans, R. (1969). *J. Radioanal. Chem.* **3**, 99.

Cooper, J. A., Rancitelli, L. A., Perkins, R. W., Haller, W. A., and Jackson, A. L. (1969). *In* "Modern Trends In Activation Analysis" (J. R. Devoe, ed.), Vol. 2, p. 1054. NBS Special Publ. 312, Washington, D.C.

Filby, R. H., Davis, A. I., Wainscott, G. G., Haller, W. A., and Cassatt, W. A. (1969). Gamma-Ray Energy Tables For Neutron Activation Analysis. Washington State Univ. Rep. No. WSUNC-97.

Graber, F. M., Lukens, H. R., and MacKenzie, J. K. (1970). *J. Radioanal. Chem.* **4**, 229.

Heath, R. L. (1964). Scintillation Spectrometry. Scintillation Gamma-Ray Spectrum Catalogue. Phillips Petroleum Co. AEC Res. and Develop. Rep. TID 4500, 2nd ed.

Hughes, D. J., and Schwartz, R. B. (1958). Neutron Cross Sections. Brookhaven Nat. Lab. Rep. BNL 325, Upton.

Johnson, R. A. (1969). *In* "Modern Trends In Activation Analysis" (J. R. Devoe, ed.), Vol. 2. NBS Special Publ. 312, Washington, D.C.

Kenna, B. T., and Conrad, F. J. (1966). Tabulation of Cross Sections, Q-Values, and Sensitivities for Nuclear Reactions of Nuclides with 14 MeV Neutrons. Sandia Corp. Rep. SC-RR-66-229, Albuquerque, New Mexico.

Lederer, C. M., Hollander, J. M., and Perlman, I. (1967). "Table of Isotopes." Wiley, New York.

Lukens, H. R. (1964). Estimated Photopeak Specific Activities In Reactor Irradiations. General Atomic Rep. GA-5073.

Lukens, H. R., Yule, H. P., and Guinn, V. P. (1965). *Nucl. Instrum. Methods* **33**, 273.

Lukens, H. R., Bryan, D., Hiatt, M. A., and Schlesinger, H. L. (1971). Development of Nuclear Analytical Techniques For Oil-Slick Identification. Rep. No. Gulf-RT-A10684 Gulf Energy and Environmental Systems.

Lundgren, F. A., and Nargowalla, S. A. (1968). *Anal. Chem.* **40**, 672.

Lutz, G. J., Boreni, R. J., Maddock, R. S., and Meinke, W. W. (1971). Activation Analysis: A Bibliography. NBS Tech. Note 467, in two volumes.

McCrary, J. H., Morgan, I. L., and Baggerly, L. I. (1961). *Proc. 1961 Int. Conf. Modern Trends Activation Analysis*, p. 24.

Peteu, G. (1969). Nuclear Techniques and Mineral Resources, p. 383. Int. Atomic Energy Agency.

Roy, J. C., and Hawton, J. J. (1960). Table of Estimated Cross Sections for (n, p), (n, α), and (n, 2n) Reactions in a Fission Neutron Spectrum. Atomic Energy of Canada Rep. AECL-1181, Chalk River.

Schonfeld, E., Kilby, A. H., and Davis, W., Jr. (1966). *Nucl. Instrum. Methods* **45**, 1.

CHAPTER 13

Nuclear Magnetic Resonance Spectrometry

Don Ware* and R. S. Codrington

Varian Associates
Palo Alto, California

Introduction

Since the initial demonstrations of the phenomenon of nuclear magnetic resonance (NMR) in late 1945 by Purcell and co-workers and independ-

* Present address: Bruker Magnetics, Inc., Burlington, Massachusetts.

ently by Bloch and co-workers, it has grown to be a major tool for the investigation of materials. It has been a particularly powerful aid to chemical structure identification in liquids and solutions, and its use is widespread in all other fields of analysis.

In this chapter we briefly outline the theory of the NMR process and the parameters which affect the position, amplitude, and shape of the nuclear resonance signal. In the section on apparatus we discuss the various components of an NMR spectrometer to enable the reader to choose the configuration most suited to his needs. Throughout the chapter we have leaned heavily toward procedures possible on commercially available spectrometers but have referred to some recent experiments which, although in the research stage, show great promise in materials analysis.

In the applications discussed, no attempt at completeness has been made or is remotely possible in merely one chapter. We have covered the main areas of activity and leave it to the reader's inventiveness to extend the field to his own area of interest.

1 Theory

1.1 Magnetic Properties of Nuclei

The nuclear properties which form the basis for nuclear magnetic resonance are the mass M, the charge Ze, and the angular momentum $Ih/2\pi$. The charge is in units of the charge on the electron, e, so that the atomic number Z is an integer. Similarly, the units of angular momentum chosen are $h/2\pi$, where h is Planck's constant and I is either an integer or half-integer. A model of the nucleus which is useful in explaining its magnetic properties is given in Fig. 1.

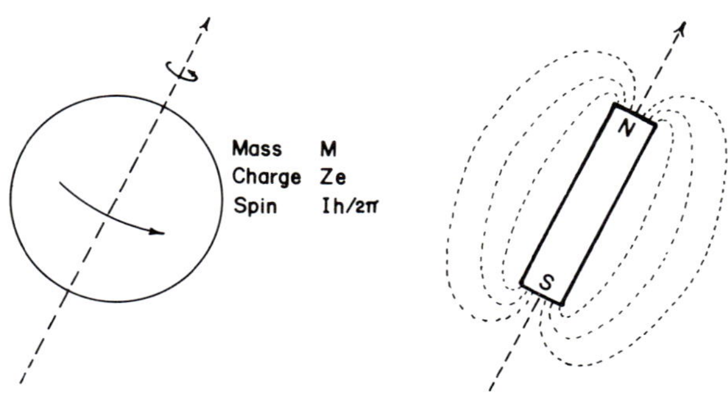

Fig. 1. The charge and spin properties of some nuclei make them behave like small magnets when placed in a magnetic field.

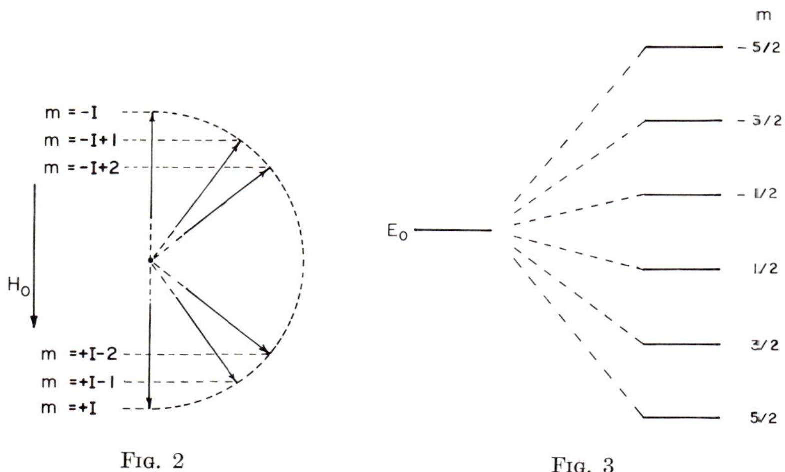

FIG. 2

FIG. 3

FIG. 2. When a nucleus with spin I is placed in a magnetic field it may be in any one of $2I + 1$ possible energy states corresponding to the projection of I in the direction of the field taking on values $I, I - 1, I - 2, \ldots, -I + 1, -I$.

FIG. 3. A nucleus with a spin $I = 5/2$ having energy E_0 in the absence of a magnetic field can exist in any one of $2I + 1 = 6$ equally spaced energy states when subjected to a field H_0. The separation of these states is proportional to H_0.

The angular momentum causes the nucleus to exhibit gyroscopic motion when subjected to a torque. The combination of the angular momentum and the charge may be visualized as a circulating current which produces the dipolar magnetic field of the nucleus specified by the dipolar moment μ.

When a nucleus possessing a nonzero spin I is placed in a magnetic field H_0, it will orient itself in one of $2I + 1$ possible orientations with respect to H_0. The projection m of the spin angular momentum I in the direction of H_0 takes on integer or half-integer values $I, I - 1, I - 2, \ldots, -I$ as represented in Fig. 2.

The energy of interaction of the magnetic moment with the applied magnetic field H_0 is

$$E_{\text{m}} = -\mu H_0 m/I \tag{1}$$

The ground state energy of any system with nuclei of nonzero spin I is therefore broken up into the $2I + 1$ states or Zeeman levels, with the energy separation of adjacent states given by

$$\Delta E = \frac{-\mu H m}{I} - \frac{-\mu H(m + 1)}{I} = \mu H/I \tag{2}$$

It should be noted that this energy difference is proportional to the applied field H_0, so that the energy level system may be pictorially represented as in Fig. 3. This is the case for a "free spin," i.e., a spin experiencing no other interactions. The modification of these levels due to other interactions will be discussed in later sections.

A magnetic nucleus may be stimulated to jump to an adjacent energy level, i.e., $\Delta m = \pm 1$, if it is irradiated with electromagnetic energy of frequency f such that, from Eq. (2),

$$\Delta E = fh = \mu H/I \tag{3}$$

Solving Eq. (3) for f,

$$f = \mu H/hI \tag{4a}$$

or

$$f = \gamma H \tag{4b}$$

where γ is a constant sometimes called the magnetogyric ratio of a particular nucleus. Equation (4b) is sometimes called the Larmor equation. The magnetogyric ratio γ usually appears in the literature as $\gamma = 2\pi\gamma$. We have chosen the modified definition so that frequencies may be expressed in hertz rather than in radians.

The nuclear dipole μ, when placed in an applied magnetic field H_0, will experience a torque which will cause it to precess about the direction of H_0 at a rate given by

$$d\boldsymbol{\mu}/dt = 2\pi\gamma(\boldsymbol{\mu} \times \mathbf{H}) \tag{5}$$

The precession frequency f will be independent of the energy state m, and can be seen to be identical with the frequency given in Eq. (4).

If, in addition to the static magnetic field H_0, we have a radio-frequency field with a vector component perpendicular to H_0, the nuclear dipole will

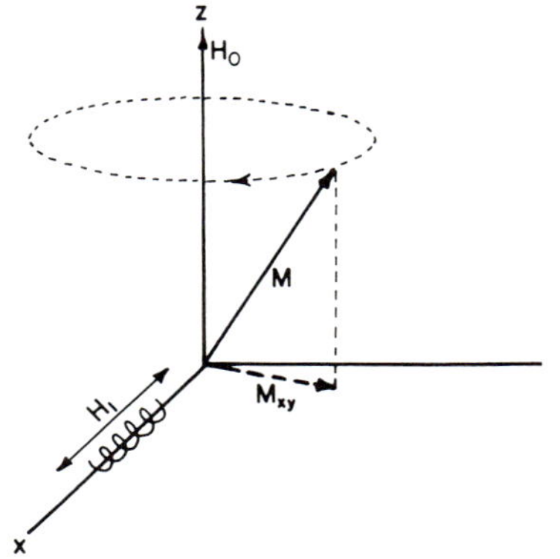

FIG. 4. Tipping of the magnetic moment M by a radio-frequency field H_1 applied perpendicular to the magnetic field H_0 induces a signal in the pickup coil proportional to M_{xy}.

experience a second torque, where the H in Eq. (5) is now the strength of the radio-frequency field. This second torque will have a net effect upon the dipole only if the radio frequency is close to the precession frequency f. This is known as the resonance condition and is shown schematically in Fig. 4.

Values of γ together with the spin I and the natural abundance for several magnetic isotopes are given in Table 1. Included in Table 1 is a column showing the sensitivity at constant magnetic field for each isotope, relative to hydrogen. These sensitivities will be discussed in greater detail in Section 2.5.

Also included in the table is a column showing the electric quadrupole moment, which will tend to complicate the energy level diagram when the nucleus is subjected to an electric field gradient. The electric quadrupole moment only occurs in isotopes with spin number greater than $\frac{1}{2}$.

TABLE 1

NUCLEAR MAGNETIC RESONANCE PROPERTIES OF SOME TYPICAL ISOTOPES

Isotope	γ MHz/T[a]	Natural abundance (%)	Relative[b] sensitivity (%)	I[c] Spin	Electric[d] quadrupole moment Q
^{1}H	42.577	99.98	100.0	1/2	
^{2}H	6.536	0.02	1.0	1	2.73×10^{-1}
^{7}Li	16.546	92.58	29.3	3/2	−3
^{9}Be	5.984	100.00	1.4	−3/2	5.2
^{10}B	4.576	19.58	2.0	3	7.4
^{11}B	13.660	80.42	16.5	3/2	3.55
^{13}C	10.706	1.11	1.6	1/2	
^{14}N	3.076	99.63	0.1	1	7.1
^{15}N	4.314	0.37	0.1	−1/2	
^{17}O	5.772	0.04	2.9	−5/2	−2.6
^{17}F	40.055	100.00	83.3	1/2	
^{23}Na	11.262	100.00	9.3	3/2	14–15
^{27}Al	11.094	100.00	20.6	5/2	14.9
^{29}Si	8.458	4.70	0.8	−1/2	
^{31}P	17.235	100.00	6.6	1/2	
^{33}S	3.266	0.76	0.2	3/2	−6.4
^{35}Cl	4.172	75.53	0.5	3/2	−7.89
^{37}Cl	3.472	24.47	0.3	3/2	−6.21

[a] 1 T = 10^4 G.

[b] Relative to ^{1}H for the same number of nuclei at constant field.

[c] In multiples of $h/2\pi$.

[d] In multiples of femtometers^2.

1.2 Thermal Equilibrium and Nuclear Magnetization

Since the population of spin states in thermal equilibrium is governed by the Boltzmann condition, the population N_m, in the energy state corresponding to the spin projection m, is given by

$$N_m \approx \frac{N}{2I+1} \exp\left(-\frac{\mu H_0 m}{IkT}\right) \approx \frac{N}{2I+1}\left(1 - \frac{\mu H_0 m}{IkT}\right) \tag{6}$$

where N is the total spin population, k is the Boltzmann constant, and T is the absolute temperature of the equilibrium spin system. The approximation above is possible because $\mu H_0 \ll kT$ in fields normally achieved with laboratory magnets except at very low temperatures (i.e., $T \approx 0°\text{K}$). If N_m in Eq. (6) is summed over all $2I + 1$ states, the total magnetization, the vector sum of the N individual spin vectors is

$$\mathbf{M} = C\mathbf{H}_0/T \tag{7}$$

where

$$C = N\mu^2 I(I + 1)/3k$$

is the Curie constant. We can see from Eq. (7) that the total magnetization is a function of the magnetic field strength and of the nuclear dipole moment. It is the effect on this total magnetization which we will monitor during nuclear magnetic resonance experiments.

1.3 The Bloch Equations

Classically, the vector magnetization **M**, when subjected to a resonant radio-frequency field, will rotate in a manner described by Eq. (5). In practice however, it has been found that two additional conditions must be applied to the behavior of **M**, which can best be described in terms of two characteristic time constants T_1 and T_2 called "relaxation time constants."

The phenomenon of nuclear magnetic resonance may generally be described by a set of differential equations developed by F. Bloch. These equations in addition to the conditions on the magnetization **M** imposed by Eq. (5), include two conditions which describe the establishment or loss of equilibrium.

The first assumption of Bloch was that when a sample is initially placed in a magnetic field, the magnetization **M** is zero. If M_0 is the equilibrium value in the z direction, then

$$dM_z/dt = -(M_0 - M_z)/T_1 \tag{8}$$

where T_1 is a measure of the attainment of thermal equilibrium and is called the "spin–lattice relaxation time constant."

The second assumption of Bloch was that as soon as **M** is directed in some direction other than the direction of the applied field, the component in the xy plane begins to decay exponentially according to

$$dM_x/dt = -M_x/T_2 \tag{9a}$$

$$dM_y/dt = -M_y/T_2 \tag{9b}$$

where T_2 is a measure of the loss of phase coherence in the ensemble of nuclei producing **M**, and is called the "spin–spin relaxation time constant."

The complete Bloch equations are obtained by combining Eqs. (5), (8), and (9):

$$dM_x/dt = 2\pi\gamma(\mathbf{M} \times \mathbf{H})_x - (M_x/T_2) \tag{10a}$$

$$dM_y/dt = 2\pi\gamma(\mathbf{M} \times \mathbf{H})_y - (M_y/T_2) \tag{10b}$$

$$dM_z/dt = 2\pi\gamma(\mathbf{M} \times \mathbf{H})_z - [(M_0 - M_z)/T_1] \tag{10c}$$

If the vector field **H** is made up of the static field H_0 in the z direction and an oscillating field $2H_1 \sin 2\pi ft$ in the x direction, Eqs. (10) have an equilibrium solution corresponding to M_z and the components of the magnetization vector, u and v, which are respectively in phase and of orthogonal phase with the driving radio-frequency field. The solutions are

$$u = 2\pi\gamma M_0 H_1 T_2 \frac{2\pi(f_0 - f)T_2}{1 + (2\pi\gamma H_1)^2 T_1 T_2 + (f_0 - f)^2(2\pi T_2)^2} \tag{11a}$$

$$v = 2\pi\gamma M_0 H_1 T_2 \frac{1}{1 + (2\pi\gamma H_1)^2 T_1 T_2 + (f_0 - f)^2(2\pi T_2)^2} \tag{11b}$$

$$M_z = M_0 \frac{1 + (f_0 - f)^2(2\pi T_2)^2}{1 + (2\pi\gamma H_1)^2 T_1 T_2 + (f_0 - f)^2(2\pi T_2)^2} \tag{11c}$$

The signal corresponding to the u magnetization is frequently called the dispersion signal, while the signal corresponding to the v magnetization is the absorption signal. The shapes of these signals are shown in Figs. 5a and b.

If the intensity of the radio-frequency field H_1 is set to a low value, the term in the denominator, $(2\pi\gamma H_1)^2 T_1 T_2$, may be ignored. The v-mode signal [Eq. (11b)], will have a maximum value when $f = f_0$ and will be half the maximum value when

$$4\pi^2(f_0 - f)^2 T_1 T_2 = 1 \tag{12}$$

The line width at half amplitude of the v-mode resonance is related to T_2 by the simple equation

$$\Delta f = 1/\pi T_2 \tag{13a}$$

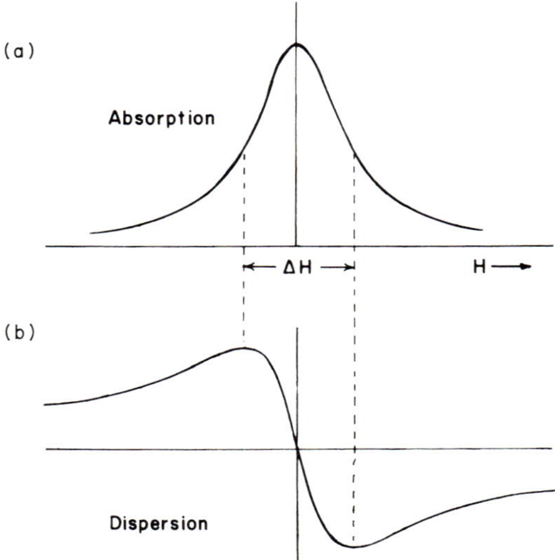

FIG. 5. The dispersion and absorption mode signals corresponding to the u- and v-mode solutions to the Bloch equations. The line width ΔH plays an important role in analyzing complex spectra.

or in terms of magnetic field

$$\Delta H = 1/\pi\gamma T_2 \tag{13b}$$

for a Lorentzian line, a line with shape given by the general equation $y = a/(1 + x^2)$.

1.4 Saturation

When excited by a radio-frequency field H_1, the probability of a nuclear spin making an energy transition $\Delta m = +1$ is the same as its probability of making the opposite transition $\Delta m = -1$. However, owing to the excess of spins in the lower energy levels dictated by the Boltzmann distribution, Eq. (6), there will be a net absorption of energy. The spin excess in the lower levels will thus tend to approach zero. This equalizing of the populations, or saturation, is opposed by the spin–lattice relaxation process, with its characteristic time constant T_1. Therefore, the degree of saturation one may obtain for a given H_1 will be proportional to T_1.

If the nuclear system is in equilibrium with the radio-frequency field and the lattice, its population distribution may still be governed by a Boltzmann distribution, which will be different from the distribution in the absence of H_1. Equation (6) can be solved for a new temperature T_s which we call the "spin temperature." We can produce spin temperatures warmer or colder than the laboratory temperature, and at complete saturation, $T_s = \infty$.

The Bloch equations given in the previous section are applicable only when the spin system is very close to its lattice temperature equilibrium value. This dictates that we must keep the term $(2\pi\gamma H_1)^2 T_1 T_2$ small during nuclear observations.

1.5 Line Widths and Multiple Line Structure

The magnetic environment of an individual nucleus may be influenced by several factors. In the following subsections, several of these factors are discussed in terms of how NMR signals may be used to study fundamental properties of materials, including molecular motion and molecular structure.

1.5.1 *Dipole–Dipole Interaction*

In a solid material where the nuclei are static, a given nucleus will be influenced by the dipole magnetic fields of its neighbors. The energy of interaction of two dipoles separated by a distance $\mathbf{r}$ is given by

$$E = \frac{\boldsymbol{\mu}_1 \cdot \boldsymbol{\mu}_2}{r^3} - \frac{3(\boldsymbol{\mu}_1 \cdot \mathbf{r})(\boldsymbol{\mu}_2 \cdot \mathbf{r})}{r^5} \tag{14}$$

In terms of the effect on the line width of the NMR signal, Eq. (14) may be approximated by assuming that μ_2 produces a magnetic field at the site of μ_1 of intensity

$$H_d = \tfrac{3}{2}\mu_2 r^{-3}(3\cos^2\theta - 1) \tag{15}$$

where θ is the angle between the internuclear vector and H_0.

In solids and liquids, where internuclear distances are approximately 10^{-1} nm or 1 Å, the magnitude of the dipolar field H_d will be about 10^{-3} T. Since in a solid the principal contribution to the line width of an NMR signal results from the dipolar interaction with close neighbors, typical line widths in solids will fall in the range 10^{-4}–10^{-1} T.

The line shapes in solids tend not to have the Lorentzian shape predicted by the Bloch equations, but are more nearly Gaussian. In treating solids therefore it is more common to use a statistical mechanical approach, using such techniques as the density matrix.

The most often used parameter of a resonance signal from a solid is the "second moment." The second moment is a function of the shape of the resonance line and can be related to the structure of the solid material containing the nuclei. For a rigid single crystal, the second moment is

$$\mathrm{SM} = \frac{3}{2}\frac{I(I+1)}{I}\mu^2 N^{-1}\sum_{j<k} r_{jk}^{-6}(3\cos^2\theta_{jk} - 1)^2 \tag{16}$$

where the summation is carried out over all pairs of resonant nuclei in the lattice. If other nonresonant spin species are present, additional terms must be added to Eq. (16):

$$\frac{1}{3}\frac{I'(I'+1)}{I'}\mu^2 N^{-1}\sum_{jj'} r_{jj'}^{-6}(3\cos^2\theta_{jj'} - 1)^2 \tag{17}$$

where I' and μ' are the spin and dipole moment of the nonresonant species.

If the sample is a powder, the angular terms above may be averaged over a sphere giving a value of 4/5.

1.5.2 *Motional Narrowing*

The broad resonance lines observed in solids are not normally observed in liquids because the rapid motion of the molecules tends to average out the dipolar interaction [Eqs. (14) and (15)].

These modes of motion are thermally excited and only when the temperature is high enough will the motion be sufficient to affect the line shape. This can happen well below the "melting point" of the sample. On the other hand, the rapidly moving molecular groups will produce fluctuating electromagnetic fields at the site of a given nucleus. These fluctuations may have frequency components at the Larmor frequency and these can induce energy level transitions, as will be explained in Section 1.6.1.

The line width in liquids is determined principally by the probability of induced transitions caused by molecular fluctuation. This probability is just the inverse of the spin–lattice relaxation time constant, so that NMR line widths in liquids are given by

$$\Delta f = 1/\pi T_1 \tag{18}$$

It should be noted that the line width in liquids is usually specified in frequency units. In high resolution NMR instruments the observed line width will typically range from 0.01 to 10Hz, and the line is usually Lorentzian in shape. In some cases, where rapid spin–lattice relaxation takes place, the line width in liquids may reach 10^{-2}–10^{-1} T.

The behavior of a material as a function of temperature can be analyzed by measuring the NMR relaxation time constants at different temperatures. The general behavior of T_1 and T_2 is shown in Fig. 6. The gradual reduction of T_2 shown in this figure is typical of the behavior in complex molecules such as polymers and biochemicals. In simpler molecules the

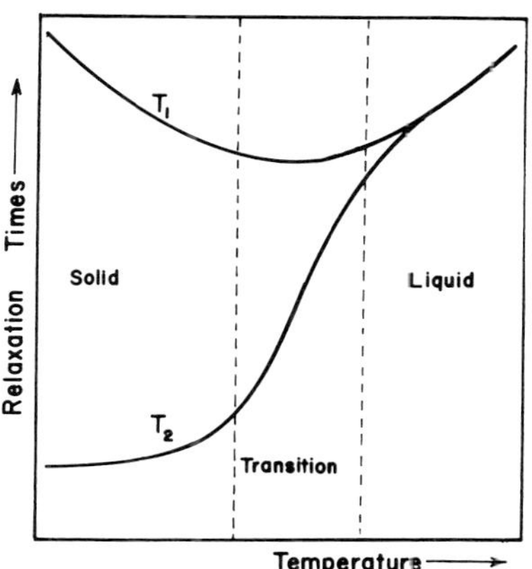

Fig. 6. The relaxation time constants T_1 and T_2 vary widely. In liquids $T_1 \approx T_2$. In solids T_1 may be seconds or hours while T_2 may be of the order of 10 to 100 μsec.

reduction in T_2 would be more abrupt because the simpler structure would have all of its modes of motion excited at similar temperatures.

1.5.3 *Chemical Shift and Molecular Structure*

The very narrow NMR line widths observed in liquids (and gases) permit other types of intra- and intermolecular magnetic interactions to be observed which are normally obscured in solids by the broad lines. The most important of these is the shift in the resonance frequency of nuclei in different chemical groups caused by the tendency of the surrounding electrons to modify their motion in such a way as to produce a magnetic field opposing any applied field H_0. This diamagnetic field δH is a function of the electron density at the site of the particular nucleus considered and will be different for different chemical groups. There is also a paramagnetic term from "outside" electrons. The actual magnetic field at the nucleus i is therefore

$$H_i = H_0 - \delta H_i \tag{19}$$

As an example, the hydrogen NMR spectrum of ethyl alcohol, CH_3CH_2OH, consists of three lines with intensities in the ratio 3:2:1. In this molecule the CH_2—O bond tends to be polar so that the electron density is attracted from the OH hydrogen to the oxygen and tends to concentrate in the CH_3 group. As a result the OH hydrogen experiences the lowest diamagnetic field, and the CH_3 group the strongest. In plotting the spectrum as a function of field (Fig. 7), the OH resonance occurs at the lowest applied field and the CH_3 resonance at the highest field. In this particular spectrum, the line widths of the three resonances are all equal and are determined by the applied magnetic field inhomogeneity. The general solutions of the Bloch equations, Eq. (11), show that the area

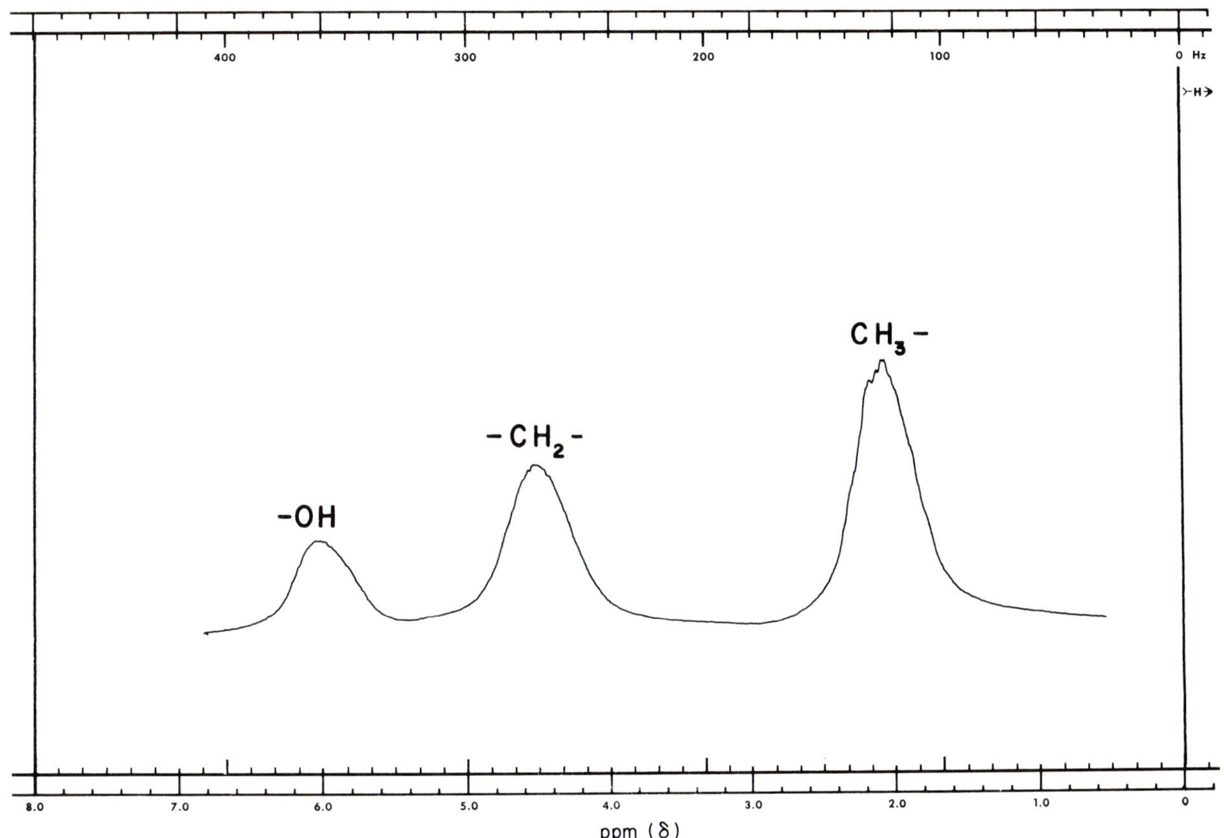

Fig. 7. The hydrogen spectrum of ethyl alcohol (CH_3CH_2OH), observed with moderate resolution, consists of three chemically shifted resonance lines with areas in the ratio of 1:2:3.

under a resonance curve is proportional to the number of nuclei. We may therefore use an integrator to obtain quantitative population ratios from the NMR spectrum.

The diamagnetic field δH produced at a given nuclear site is proportional to the applied field H_0. As a result the separation of the component lines in a chemically shifted spectrum is also proportional to the applied field and it is desirable to observe such a spectrum in the highest field available.

1.5.4 *Scalar Spin Coupling*

The spin I of a nucleus can interact with the spin S of an electron situated momentarily at the site of the nucleus in such a way that the spin state of the nucleus will be transmitted by the electron, possibly via other electrons, to another nucleus. A scalar coupling **J** can therefore exist between two nuclei in a given molecule, even though the molecule is undergoing rapid motion. This coupling is field independent and independent of molecular orientation.

In the ethyl alcohol molecule (see Fig. 7), the OH group is usually exchanging so rapidly with OH groups on other molecules and with water, which is usually present, that the OH hydrogen does not normally interact with the other hydrogens. On the other hand, the CH_3 and CH_2 hydrogens can interact through the mechanism of the $I_1 \cdot S$, $S \cdot I_2$ interaction.

The resulting $I_1 \cdot I_2$ interaction causes the ensemble of CH_3 hydrogens to experience three possible energy contributions from the two CH_2 hydrogens, corresponding to the three different spin states, i.e., $\upharpoonleft\upharpoonleft$, $\upharpoonleft\downharpoonright$ or $\downharpoonleft\upharpoonright$, and $\downharpoonleft\downharpoonright$, where the direction of the arrow means alignment of the spin with or against the applied field H_0. The probability of these spin states occurring is in the ratio of 1:2:1 and consequently the CH_3 resonance is split into a triplet. The CH_2 hydrogens similarly experience four energy states with a probability ratio 1:3:3:1 and their resonance is split into a quartet. The number of energy states is given by

$$n = 2NI + 1$$

where N is the number of nuclei of spin I being interacted with.

The complete spectrum of ethyl alcohol showing the effect of spin coupling is given in Fig. 8.

1.5.5 *The Quadrupole Interaction*

Nuclei with spin quantum number I greater than $\frac{1}{2}$ possess an electric quadrupole moment Q resulting from the deviation of the nuclear charge distribution from spherical symmetry. This quadrupole moment will interact with electric field gradients $\partial^2 v/\partial z^2$ at the site of the nucleus causing a

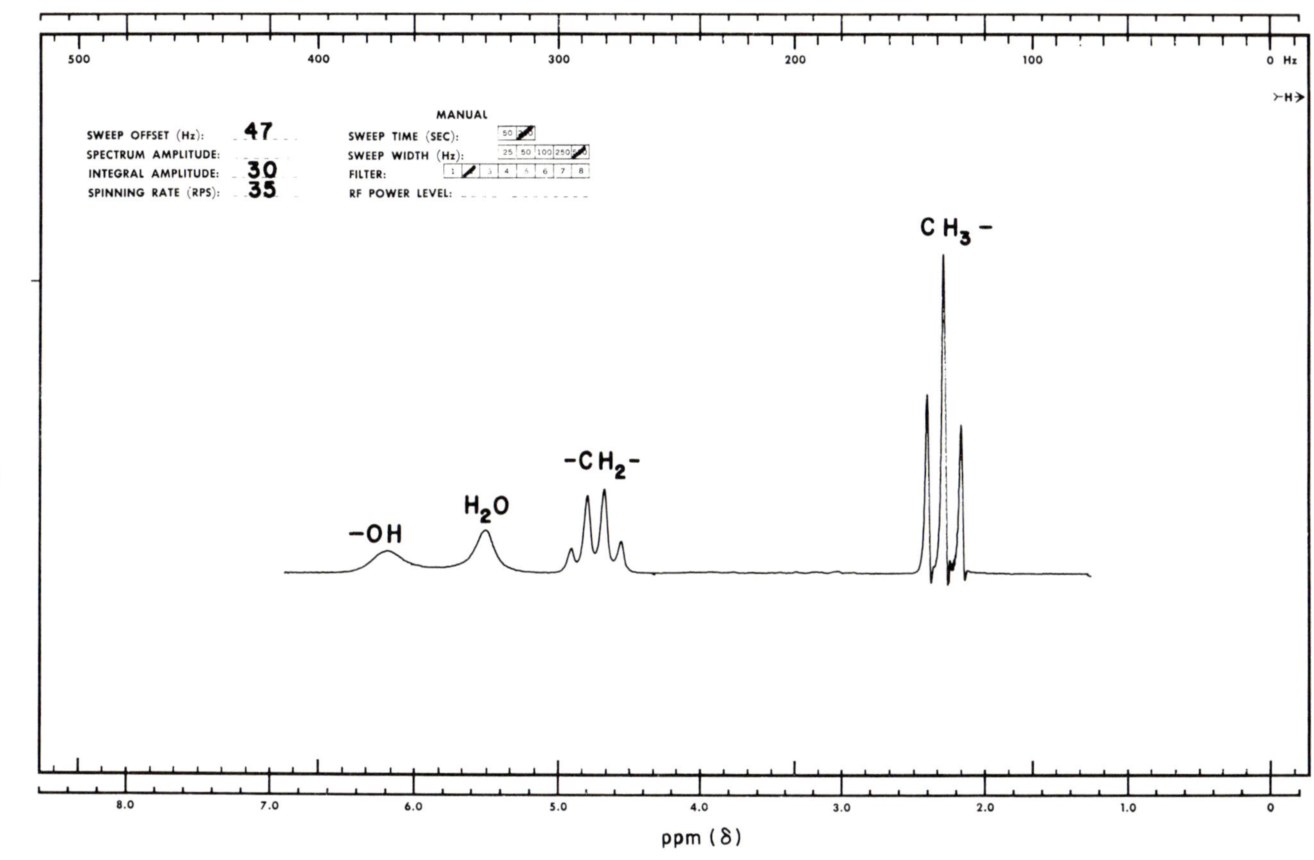

FIG. 8. The hydrogen spectrum of ethyl alcohol observed with a high-resolution spectrometer. This spectrum shows the effect of spin coupling between the CH_2 and CH_3 hydrogens. Coupling to the OH hydrogen was removed by adding water to the sample.

perturbation in the magnetic energy levels. We will thus see $2I$ separate lines at energy levels given to first order by

$$E_m = \frac{-m\mu H_0}{I} + \frac{eQ}{4I(2I-1)}[3m^2 - I(I+1)]\frac{\partial^2 V}{\partial z^2} \tag{20}$$

where the first term is the perturbation due to the nuclear magnetic interaction given in Eq. (2) and the second term is due to the quadrupole interaction.

For nuclei experiencing an electric field gradient of axial symmetry,

$$\partial^2 V/\partial z^2 = \tfrac{1}{2}eq(3\cos^2\theta - 1) \tag{21}$$

where θ is the angle between the symmetry axis and H_0, e is the electronic charge, and

$$eq = \int \rho(3\cos^2\theta - 1)r^{-3}\,dV$$

where ρ is the charge density. The integral is taken over all charges outside the nucleus. The vector $\mathbf{r}$ joins the nucleus to the volume element dV at an angle θ with the symmetry axis. We therefore find resonant lines at frequencies

$$f_{m\to m-1} = f_0 + \frac{3e^2Qq(2m-1)}{8I(2I-1)h}(3\cos^2\theta - 1) \tag{22}$$

For nuclei with small quadrupolar interactions such as ^{23}Na in $NaNO_3$, this formula gives good results, but calculations to higher orders are necessary for nuclei with large quadrupolar interactions such as ^{27}Al in Al_2O_3. If the quadrupole interaction is strong enough, we may observe a transition in the absence of an external magnetic field. This technique, requiring frequency sweep covering the range 1–1000 MHz, is commonly known as "pure nuclear quadrupole resonance."

1.5.6 *Knight Shift*

In metals, some semiconductors, and ferromagnetic materials, the resonance field for a given frequency is always higher than for the same nucleus in a diamagnetic specimen. This shift, discovered by Knight, is due to interaction of the nuclei with magnetic fields produced at their site by conduction electrons near the top of the Fermi distribution, i.e., electrons in partially filled energy levels. The shift is field dependent, $\Delta H/H_0$ being between 10^{-5} and 10^{-2}. It may also show anisotropy if the sample is noncubic or has lattice defects.

1.6 Spin–Lattice Interactions

In order that the spin system and the lattice may come into thermal equilibrium, they must be able to exchange energy. There are three basic mechanisms of energy exchange, and we can use measurements of the relaxation time constant T_1 to probe these mechanisms and thus obtain information about the lattice.

1.6.1 *Molecular Reorientation and Diffusion*

This is the most important mechanism for relaxation in liquids and gases, and may be an important mechanism in solids such as polymers and crystals containing organic molecules which are free to rotate wholly or in part or to diffuse. The local field seen at a nuclear site due to the neighboring spin moments will be a rapidly fluctuating function of time due to the relative motion. If this time fluctuating field has a Fourier component at the resonant frequency f_0, it can induce spin transitions in the same way as a radio-frequency field, except that it cannot bring the spin temperature beyond the lattice temperature as the fluctuating field is due to a thermal reservoir or "phonon bath" which has a definite temperature.

It is possible to relate the motion in a sample to the value of T_1 through the Debye correlation time τ_c, which is directly proportional to the viscosity η.

$$\frac{1}{T_1} = \frac{C_1\tau_c}{1 + 4\pi^2 f_0^2 \tau_c^2} + \frac{C_2\tau_c}{1 + 16\pi^2 f_0^2 \tau_c^2} \tag{23}$$

where C_1 and C_2 are terms containing the spectral density function of the motion, and for a molecule of effective radius a,

$$\tau_c = 4\pi\eta a^3/3kT$$

For relatively nonviscous liquids, $2\pi f_0 \tau_c \ll 1$, giving an inverse relationship between T_1 and viscosity.

In solids with rotatable groups, such as the CH_3 groups in xylene, the correlation time may be related to the potential energy barrier V hindering rotation.

$$\tau_c = \tau_0 \exp(V/RT) \tag{24}$$

where τ_0 is a constant, R is the gas constant, and T the absolute temperature. A plot of T_1 versus T may show several minima as the phonon bath temperature selectively excites rotational modes corresponding to the resonance frequency.

If diffusion is taking place in the solid lattice, this will constitute a phonon bath of very low frequency. Energy exchange between this bath and the spin system can take place if the Zeeman energy is low enough or if the spin system can be brought into a quasi equilibrium state with closely spaced energy levels.

1.6.2 *Paramagnetic Impurities*

For rigid lattices such as are found in ionic crystals, lattice vibrations are a very poor means of thermal relaxation if the nuclear spin species does not have a quadrupole moment. In these crystals relaxation takes place via paramagnetic impurities, even when these impurities are very dilute. The paramagnetic impurities are able to couple strongly with the lattice vibrations and the nuclei couple to the distant impurities via spin–spin interactions with intervening nuclei. This allows spin energy to "diffuse" toward the impurity sites. If the sample contains many paramagnetic ions, T_1 may be very short, imparting a broadening to the resonance line shape. This broadening will also be seen in liquids doped with paramagnetic ions.

1.6.3 *Quadrupole Relaxation*

The electric quadrupole moment of a nucleus in a molecule undergoing motion will couple with the fluctuating local electric field gradients and induce relaxation in the same way as for the dipole moment. This mechanism may be very effective in relaxing the nucleus, especially if the quadrupole moment is large. The very short T_1's resulting may give the resonance line a very broad Lorentzian shape, up to 10^{-1} T in width.

In solids containing nuclei with quadrupole moments, lattice vibrations will produce Fourier components of the electric field gradient at frequencies equal to f_0 and $2f_0$ which will induce $\Delta m = \pm 1$ and $\Delta m = \pm 2$ transitions. These interactions, in contrast to the lattice produced magnetic fluctuations, are able to produce rapid relaxation of the nuclear spin energy. The relaxation time constant will be inversely proportional to the quadrupole moment.

1.7 Double Resonance

In nuclear magnetic resonance spectroscopy, it is often desirable to irradiate the sample with more than one radio-frequency source at the same time. This second source, the "pump," is often much stronger than the first source, the "observer." The purpose of the pump is to saturate a nuclear (or sometimes an electron) transition and thereby remove or modify its effect upon the observed transition.

1.7.1 *Spin Decoupling*

The interpretation of high-resolution NMR spectra frequently involves decisions concerning the possible ways nuclei are coupled to produce the observed spin multiplet patterns. The spectrum of ethyl alcohol in Fig. 8 is relatively easy to interpret because the chemical shifts are large compared to the spin–spin coupling. This is not always the case however and in some spectra the multiplets from different chemical groups overlap.

Spin decoupling is a useful technique for sorting out the coupling of the different groups. The pump is used to saturate a particular region in the frequency spectrum while the remainder of the NMR spectrum is being observed. Those energy levels separated by the decoupling frequency will be saturated and interactions between nuclei in those states and other nuclei will be effectively removed. As a result the multiplets in other groups caused by the interaction with the irradiated group collapse into a single spectral line.

In Fig. 9, an example of spin decoupling in ethyl alcohol is shown. With the decoupling frequency set to the resonance frequency of the CH_3 hydrogens, the CH_2 hydrogen spectrum is collapsed into a single line. Simi-

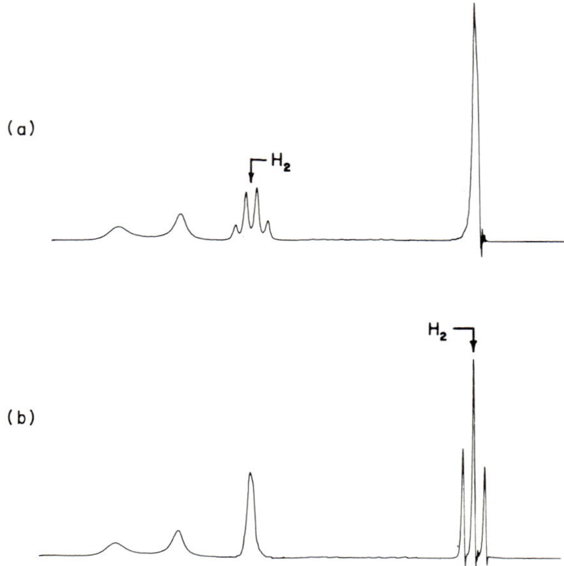

FIG. 9. The spin multiplets of one hydrogen group may be collapsed by irradiating the hydrogens of the group to which they are coupled with a strong radio-frequency field H_2. In Fig. 9a the CH_2 group was irradiated while in Fig. 9b the CH_3 group was irradiated.

larly, when the decoupler is set to the CH_2 resonant frequency, the CH_3 triplet collapses to a single line.

1.7.2 *Noise Decoupling*

In some instances it is desirable to decouple all spectral components in a broad frequency range. This can be done by modulating the single frequency from the decoupler transmitter with a Gaussian noise or pseudorandom noise so that the output of the modulator covers the desired frequency range around the decoupler frequency. Noise decoupling is particularly useful when observing a weak signal from a nucleus such as ^{13}C which is coupled to many hydrogens in a complex molecule. The ^{13}C spectrum, which would normally be spread out over many spin multiplets, can be collapsed into a few chemically shifted lines.

1.7.3 *Overhauser Effect and Dynamic Nuclear Polarization*

If a sample contains two (or more) spin species, either nuclei or nuclei and free electrons, it is possible to change the population distribution of one species by pumping the other species with the proper radio frequency. This pumping will lead to either an enhancement or a deenhancement of the observed resonance. The amount and the sign of the signal amplitude change will depend upon the coupling between the observed and the pumped species, the difference in their γ's, and the amount of pumping power used. The pumped species is often a free electron, introduced by paramagnetic doping or naturally present, and the amount of enhancement may be significant ($\sim\times 700$). The degree of enhancement will be less if the pumped species is another nucleus, but the experiment may be more easily performed since a microwave source and cavity need not be used and the second nuclear species may be naturally present and strongly coupled.

1.7.4 *Cross Relaxation*

A nuclear spin system comes to equilibrium with the lattice by exchanging energy with it or relaxing. This energy exchange will have a net direction from the "hot" to the "cold" system. In much the same manner, two spin systems, usually different nuclear species, can exchange energy and come to a mutual equilibrium. This energy exchange process is called "cross relaxation." Cross relaxation is generally promoted by supplying one or more radio-frequency sources so adjusted that the effective energy level splitting in the two systems is identical. In this way a transition in one system may be accompanied by an opposite transition in the other system, conserving the total spin energy in the two systems.

If one system has a low abundance with a low γ so that direct observation is difficult, it may be made to cross-relax with an abundant, easily observed system which can be used to monitor the effect. In this way information about the chemical and crystallographic state of the low abundance species is possible.

The effective signal from a system with a low γ can be enhanced if it can be made to cross-relax with a system with a high γ in such a manner that the low γ system is "cooled." Spins are "pumped" from the high-energy state to the low-energy state so that the total magnetization given by Eq. (7) is increased. The high γ system in these experiments is usually an electron spin system, and specific techniques known as the "Overhauser effect" and "dynamic nuclear polarization" have been widely used in NMR.

1.8 Continuous Wave and Transient Techniques

In cw NMR the radio frequency or the magnetic field is scanned over the resonant region of interest in a time period dictated by the relaxation time constants T_1 and T_2. The recorded spectrum is in the frequency domain; i.e., resonance strength is recorded as a function of frequency (or field).

If, instead of irradiating the sample with a continuous frequency, we irradiate it with a very short high-intensity pulse of radio-frequency energy near resonance, the nuclei will experience a torque which will cause the microscopic magnetization vector **M** to rotate toward the xy plane. The short radio-frequency pulse contains a whole spectrum of frequencies (its Fourier transform), which cover the resonant frequencies of all the nuclei resonant near the center frequency of the pulse. If the radio-frequency field is stronger than the internuclear interactions in the sample, all the nuclei will be rotated together. If the nuclei are allowed to precess in the applied radio-frequency field through $\pi/2$ rad, this is called a "$\pi/2$-pulse," whereas a pulse which causes the spins to rotate through π rad is called a "π-pulse."

After the pulse is shut off, the individual nuclei will dephase in time with a time constant T_2 due to their different precession frequencies in the constant magnetic field H_0 plus local fields ΔH_0 due to magnet inhomogeneities and other magnetic dipoles, and the signal induced into the receiver coil will decay. This signal, the Bloch decay, is the Fourier transform of the cw absorption signal, and is in the time domain, Fig. 10. If we take the time-to-frequency Fourier transform in an analog or digital device, we can recover the frequency domain spectrum. The nuclei in the xy plane will return to the equilibrium z direction with a time constant T_1, and the en-

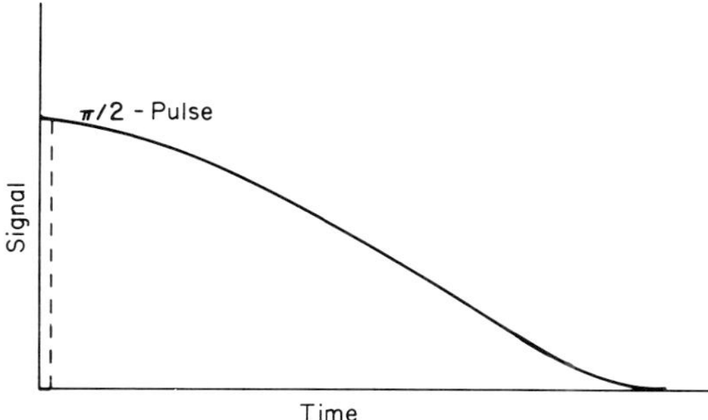

FIG. 10. The Bloch decay following a $\pi/2$-pulse of radio-frequency power. The signal is induced into a coil mounted with its axis in the xy plane. The Bloch decay is the Fourier transform of the nuclear absorption signal, Fig. 5.

tire transient experiment may be repeated after sufficient nuclei have returned so that a usable signal is produced after a second pulse.

The amplitude of the transient signal immediately after the first radio-frequency pulse is directly proportional to the equilibrium magnetization M_0, independent of T_1 and T_2 and so, from Eq. (7), we may use this as a quantitative analysis tool. If we measure the amplitude of the transient following a second pulse, we may fit this value into Eq. (8) to calculate a value for T_1, the spin–lattice relaxation time constant. If the Fourier transform of the entire decay is made, the T_1's of individual lines in a high-resolution spectrum may be measured

In principle, any measurement possible by cw techniques can be made with transient techniques, and a large number of measurements are only possible using transient techniques. Order of magnitude improvements in the signal-to-noise ratio for a given experimental time are possible using transient or Fourier transform techniques in high-resolution NMR.

2 Instrumentation

The instrumentation of NMR is basically quite simple. One needs a source of radio-frequency power, the transmitter, which can be at the milliwatt level for cw experiments and as high as tens of kilowatts for transient experiments. One also needs a means of irradiating the sample with the radio-frequency power. This usually takes the form of a coil or coils of wire into which the sample is placed. The means of detecting changes in the sample due to resonant absorption of the radio-frequency power can be another coil (cross-coil probe), or the same coil as is used for the transmitter (single-coil probe). A receiver is used to amplify the signal resulting from resonance absorption and the result is detected and displayed on a recorder or oscilloscope, or the detected signal can be processed and displayed digitally. For transient experiments a programmer is needed to provide the necessary pulse sequences.

In those types of NMR experiments where a laboratory magnet is required to produce the static magnetic field, the probe containing the radiofrequency coil (or coils) is placed in a homogeneous region of this field and the resonance condition is found by adjusting either the radio frequency or the magnetic field, or both.

2.1 Magnets

The magnets used for NMR, although of bulky appearance, are of necessity precision instruments since they must be capable of producing and maintaining a magnetic field homogeneity over the sample volume on the order of one part in 10^9. Three types of magnet are commonly used for NMR; permanent magnets, electromagnets, and superconducting solenoids.

2.1.1 *Permanent Magnets*

The permanent magnets use Alnico to generate the magnetic flux and have the advantage that they require little power to operate once they are charged to the desired field. Permanent magnets are usually limited to fields of less than 2 T. Owing to the large temperature coefficient of Alnico, permanent magnets used in NMR are usually kept in a thermally regulated environment. Their main disadvantage is the inability to change the field value by more than small amounts.

2.1.2 *Electromagnets*

The electromagnets used in NMR can operate up to about 2.4 T (100 MHz for hydrogen resonance), even though the yoke, pole pieces, and pole cap materials (soft steels) begin to saturate magnetically at about 2.0 T. The coils of an electromagnet are usually driven by a well-regulated power supply and are water cooled for stability.

The size of an NMR magnet is usually specified by the diameter of the pole piece and commercial units run from about 10 to 50 cm. A pole cap diameter to air gap ratio of at least 4 or 5 to 1 is required to obtain the necessary homogeneity in iron magnets.

2.1.3 *Superconducting Solenoids*

The superconducting solenoid is usually made of niobium–zirconium or niobium–titanium wire and operates in a liquid helium Dewar system. The wire is in the superconducting state and when the ends are joined and the external current source removed, the desired current will continue to flow in the coil for as long as the wire remains superconducting. Superconducting solenoids can reach much higher fields than iron magnets (up to about 7.5 T) and are very stable when operated in the persistent mode (ends joined and power supply removed).

In addition to relying on the basic homogeneity of the magnet, high-resolution NMR systems usually have a number of current shim coils mounted either on the probe or on the magnet pole piece. The shim coils approximate various linear and higher order field gradients and can be used to enhance the homogeneity of the magnetic field at the site of the sample. A further refinement in high-resolution systems is that the cylindrical sample with its axis in the y direction is usually spun (typically 30–60 Hz) about this axis to average out x and z gradients.

If field sweep experiments are to be performed, a means of sweeping the field, either by varying the current in the magnet coils or by applying a current to additional coils, must be provided. The field sweep control is often coupled to the recorder x-axis drive so that a reproducible plot of signal as a function of field may be obtained.

2.2 Probes

The purpose of the probe in the NMR experiment is usually twofold. It must hold the radio-frequency coil (or coils) and the sample in the magnetic field, and it must provide some means for isolating the transmitter from the receiver. In some experimental arrangements, this latter function is not required. The probe may also hold coils for providing magnetic field sweep and modulation, field homogeneity adjustment, and second radio-frequency irradiation.

2.2.1 *Single-Coil*

Single-coil probes use one coil for both the transmitter and receiver. This coil, usually a solenoid, is mounted with its axis perpendicular to the magnetic field. The coil is made a part of a tuned circuit and can be coupled directly to the transmitter and receiver, or more usually it is part of a bridge circuit which allows transmitter–receiver isolation.

2.2.2 *Cross-Coil*

The cross-coil probe uses two coils mounted orthogonally to each other and to the magnetic field. The transmitter coil usually is a Helmholtz pair of large dimension, with the receiver coil inside. Both coils are made parts of separate tuned circuits, and are coupled separately to the transmitter and receiver. Table 2 gives the advantages of each type.

TABLE 2

COMPARISON OF THE ADVANTAGES OF SINGLE- AND CROSS-COIL NMR PROBES

Single-coil	Cross-coil
Compact	High transmitter/receiver decoupling
High radio-frequency efficiency	High radio-frequency homogeneity
Only two leads to probe	Coils matched independently to the transmitter and receiver circuitry
Able to do pure NQR	High sensitivity for equivalent receiver coil geometry
Able to rotate magnet around coil for single crystal experiments	

2.3 TRANSMITTER/RECEIVER AND AMPLIFIERS

The electronic circuitry needed to provide the radio-frequency field and to detect the resonance condition can be divided into three groups: the transmitter, the receiver, and the audio amplifiers.

2.3.1 *Transmitter*

The transmitter usually operates in the frequency range of 2 to 100 MHz but for some experiments it may operate at lower frequencies or at higher frequencies up to and above 300 MHz. The transmitter consists of a frequency source (or sources), and one or more stages of power amplification. If transient experiments are to be performed, a gate capable of giving better than 100 dB of attenuation is also needed.

a. The frequency source may be a tunable oscillator if high-frequency stability is not required. This oscillator is usually temperature stabilized and may tune over wide frequency ranges, usually in steps. Where frequency variation is not required, a crystal oscillator may be used. This may be a temperature-compensated crystal oscillator (TCXO); for ultimate frequency stability an oven-controlled crystal oscillator is used. In cases where it is necessary to operate with high-frequency stability and still have the ability to vary the frequency, a frequency synthesizer is used. The frequency steps on the synthesizer are dictated by the nature of the experiment, and may be as fine at 0.01 Hz or as coarse as 1 MHz. Where only a few specific frequencies are required with fine division around them, a set of fixed crystals and a mixer may be added to a synthesizer of narrow range with fine frequency increments.

The buffered oscillator output may be amplified directly, or mixed with other frequency sources and/or multiplied to some higher frequency. It may also be necessary to split the frequency into several phase channels.

In all cases, especially for high-resolution experiments, attention must be paid to spectral and phase purity.

b. The type of power amplifier used depends upon the experiments to be performed. For cw experiments, linear narrow-band tuned amplifiers are usually employed. This helps to ensure good spectral purity. The final output power can be in the milliwatt to 10-W level. For transient experiments, a gated amplifier is necessary. The gate is usually at the low power end, soon after the frequency source, but may precede the multipliers and/or mixers. The bandwidth following the gate should be wide enough to allow a fairly sharp radio-frequency pulse to be amplified. Pulse rise and fall times as fast as 0.1 μsec are necessary in some cases. For many transient experiments, Class C amplifiers may be employed, but for higher versatility a linear amplifier is desirable. Peak power outputs to the transmitter coil as high as 10–20 kW are sometimes desirable. In both cw and transient experiments, some output radio-frequency level control is needed.

2.3.2 *Receiver*

The receiver usually consists of a low-noise preamplifier followed by several voltage amplification stages. One or more local oscillators and i.f. stages may also be employed depending upon the gain required. Gains of 10^3 to 10^6 are common. The receiver system must be linear, as in many experiments the signal amplitude is the measurement of interest.

a. A low-noise cascade configuration preamplifier is most frequently used. The gain is usually small and the preamplifier provides impedence matching to drive the transmission line to the sometimes distant receiver.

b. The radio-frequency amplifiers may all be tuned to the resonant frequency, or one or more i.f. stages may be introduced. I.f. stages are convenient if narrow banding is required, but it is also necessary to use the receiver at several frequencies. We need then only change the preamplifier and the local oscillator. Recent advances in solid-state integrated circuitry have made it possible to use broad-band amplifiers following the tuned preamplifier.

In transient experiments, the receiver must be able to recover from any overload or switched-off condition as rapidly as possible. Recovery times of <10 μsec are desirable, and for work with solids, 2–3 μsec times are often needed.

c. Following amplification, the radio-frequency voltage is detected, either by a simple diode which is usually biased to provide linearity, or by a phase-sensitive detector. The phase detector receives its reference signal from the transmitter frequency source via a variable phase shifter. If i.f. stages are used in the receiver, all local oscillators must be phase locked to the same master oscillator to provide phase coherence. After detection the signal is filtered, either with a notch filter tuned to the field modulation frequency, or with a low-pass filter adjusted to pass all the frequency components of the signal.

2.3.3 *Video Amplifiers*

Following detection and filtering, the signal may be presented directly on an oscilloscope or recorded if it is strong enough. For solids this is frequently not the case and further amplification is necessary. If the static magnetic field is modulated at a low audio frequency at an amplitude smaller than the line width, the signal following radio-frequency detection will be similarly modulated and can be further amplified with audio amplifiers tuned to the modulation frequency. The signal following phase detection and filtering will be the first derivative of the nuclear resonance signal and will be slightly distorted owing to both the audio frequency and the amplitude of the modulation. This distortion can be corrected in many cases. Gains of 10^3 to 10^4 are employed, and output filter time constants of from 10^{-2} to 10^2 sec are used.

2.4 Field Stabilizers and Controllers

Several field sensing and locking techniques are used in NMR spectroscopy to increase the field stability of permanent or electromagnets. The simplest is to use some field measuring sensor such as a Hall probe. The signal from the probe is compared to a reference signal and the difference is used to adjust the magnetic field, via additional coils or via the power supply, until the difference signal is zero. A field stability of up to one part in 10^6 can be achieved with this method. The reference signal may be adjustable and calibrated in field units, allowing easy setting of the magnetic field strength.

A second method is to place coils with large numbers of turns on the pole pieces of the magnet. A high-gain integrating amplifier is connected between the two coils so that any change in magnetic flux in the coil connected to the integrator input results in a current at the output which drives the second coil to correct the flux change. Such a device is called a flux stabilizer and can achieve relatively long-term field stabilities on the order of one part in 10^7 or 10^8.

The most precise means of achieving long-term stability is the use of a field–frequency lock. A second transmitter and receiver called the "lock channel" are used to monitor a particular resonance line. The output of the lock channel receiver is used to adjust either the two transmitter frequencies or the magnetic field so that the ratio of frequency to field keeps the spectrometer "locked" to the reference line.

The reference line may be part of the NMR signal from the sample being measured, an "internal lock" system, or it may be from a sample which is separate from the measurement sample, an "external lock." In the internal case the lock sample sees exactly the same field as the measurement sample, but the lock is destroyed when the sample is removed. Internal lock stabilities of one part in 10^{10} can be achieved with slightly poorer stabilities for external locks.

If the nuclear signal in the lock channel is from the same nuclear isotope as in the observation channel, we have a "homonuclear lock." If it is a different nuclear isotope, we have a "heteronuclear lock." The choice of lock type will depend upon the experiment being performed.

2.5 Sensitivity

For cw experiments using sinusoidal modulation, the maximum signal-to-noise voltage ratio can be shown to be

$$\frac{V_s}{V_n} = \left(\frac{\pi y \zeta N \gamma I(I+1)}{24kT}\right)\left(\frac{V_c Q f_0{}^3 T_2}{kT\beta F T_1}\right)^{1/2} \tag{25}$$

where y is a factor of the order unity which accounts for detector law and modulation effects, ζ is the filling factor of the receiver coil, N is the number of nuclei per unit volume of the sample, V_s is the volume of the receiver coil, Q is the quality factor of the receiver coil, β is the bandwidth of the receiver system, and F is its noise factor. The other terms are as defined in Section 1.3. This equation is derived with the aid of the Bloch equations, Eq. (11), with $(2\pi\gamma H_1)^2 T_1 T_2$ set equal to unity.

For the nonmodulated case, as used in high-resolution work in liquids, a fourfold sensitivity increase may be expected as no modulation sidebands are generated. For transient experiments, this signal-to-noise voltage ratio at the start of the decay becomes

$$\frac{V_s}{V_n} = \left(\frac{\pi y \zeta N \gamma I(I+1)}{3kT}\right)\left(\frac{V_c Q f_0{}^3}{kT\beta F}\right)^{1/2} \tag{26}$$

a difference of $8(T_1/T_2)^{1/2}$ over the sine-modulated cw case.

In the above expressions, we can see that the signal-to-noise ratio is proportional to $\zeta N V_c{}^{1/2}$, the number of resonant nuclei present in the sample. Therefore, within the restraint of magnet size and radio-frequency power available, the larger the sample, the better. Also, we can appreciate that NMR may be used as a quantitative tool. The signal amplitude will be proportional to $f_0^{3/2}$ if such factors as coil Q and receiver gain are kept constant. We thus wish to work at the highest frequency possible within the limits of magnetic field available. Nuclei with high γ's will yield larger signals for the same resonance frequency, and if we are limited in our maximum magnetic field, low γ nuclei may have to be observed at lower frequencies, thus further reducing the sensitivity.

The overall apparatus parameter $(Q/\beta F)^{1/2}$ dictates that we use a high-Q, narrow-band, low-noise receiver. Q may be limited by receiver recovery needs and β will be dictated by the frequency content of the signal. These two factors will work against good signal-to-noise in transient systems, but may be more than offset by the increase of $8(T_1/T_2)^{1/2}$ between Eqs. (25) and (26).

The relaxation time constant factor $(T_1/T_2)^{1/2}$ plays a rather important role in NMR. In cw experiments, a long T_1 will result in a low signal-to-noise ratio, and in transient esperiments it will limit the number of observations possible per unit time. For liquids, the line width will usually be dictated by the magnet inhomogeneity; thus improved homogeneity will not only improve the spectral resolution possible, but will also improve the signal-to-noise. In solids, T_2 will be relatively short, greatly reducing the sensitivity. In transient experiments, the signal-to-noise is not directly related to T_2, but the receiver recovery time will limit the amount of the transient decay which may be seen. Certain pulse sequences are available however which may help to improve this situation.

3 Experimental Procedures

The information we want from an NMR spectrometer may be gathered into four groups: (a) resonance frequencies and intensities; (b) line widths and shapes (spin–spin relaxation time constants); (c) spin–lattice relaxation time constants; (d) information from multiple-pulse experiments.

3.1 Resonance Frequencies and Intensities

3.1.1 *Single Resonance Lines*

Most commercially available spectrometers are not designed to sweep over large frequency ranges searching for resonance lines, but if a particular isotope is expected, a search near its resonance frequency is possible, usually by sweeping the magnetic field.

In measuring, care must be taken that large Knight shifts or quadrupole splittings are not leading to incorrect identification. The state of the sample is not too critical, provided sufficient nuclei are present to obtain a signal. For conductors, powdering may be necessary to overcome the problem of skin depth. The resonance lines for the various isotopes are far enough apart, especially at high fields, that the broad lines typical of solids usually will not mask any information.

The area under the resonance curve may be used to measure the amount of the isotope present. It is usual to calibrate the spectrometer with samples of known concentration so that amplifier gains, etc., need not be considered. If intensities are being measured with a cw spectrometer, attention must be paid to the effects of relaxation time constants. If a pulse spectrometer is used, this will usually not be necessary. It is often desirable to measure the intensities of two different components of a resonance signal, due to the solid and liquid phases, for example. Here we may make use of the different relaxation time constants to separate the components, either by selectively saturating one in a cw spectrometer or by filtering out one in a transient spectrometer.

3.1.2 *Multiple Line Spectra*

In high-resolution NMR we can resolve component lines of the nuclear resonance signal with splittings due to spin coupling and shifts due to chemical environment as small as 0.1 Hz (see Sections 1.5.3 and 1.5.4). High-resolution spectrometers are usually provided with calibrated recorders which specify the sweep width in both frequency and in parts per million; $\delta = \Delta f/f_0$ referenced to the observing frequency f_0. The frequency scale is used to specify the frequency interval or spin coupling $\mathbf{J}$ between spin multiplets, while the δ scale is useful for specifying the chemical shift between spectral lines of different chemical groups. Usually a reference compound which has a single identifiable resonance line is added to the unknown sample.

In Table 3 the chemical shifts of some typical chemical groups are given for hydrogen and carbon-13 spectra. Similar tables exist for ^{14}N, ^{17}O, ^{19}F, ^{31}P, etc. These tables are found in the books on high-resolution NMR referenced at the end of this chapter.

The interpretation consists of identifying the chemically shifted groups and confirming the identification by the spin multiplet structure. If necessary, spin decoupling (see Sections 1.7.1 and 1.7.2) may be used to further confirm the interpretation. Many different versions of computer spin simulation programs exist which can also be useful in confirming a structural identification. The ratio of the intensities of the various groups is usually measured with an electronic integrator.

TABLE 3

CHEMICAL SHIFT δ FOR VARIOUS ^{1}H AND ^{13}C GROUPS RELATIVE TO TETRAMETHYLSILANE (TMS)[a]

Hydrogen		Carbon-13	
Molecule	δ^a (ppm)	Molecule	δ^a (ppm)
Si $(CH_3)_4$	0	CH_2I_2	−61
CH_3CH_3	0.8	Si $(CH_3)_4$	0
Cyclohexane	1.5	Acetic acid (CH_3)	20
CH_3S	2.0	Acetone (CH_3)	29
Acetone	2.1	$(CH_3)_2SO$	40
CH_3Cl	3.1	CH_3OH	50
Dioxane	3.5	$CHCl_3$	78
CH_3F	4.3	Benzene	129
Cyclohexene	5.6	Acrylic acid (CH_2)	133
Analine (ring)	6.6	Acetic acid (COOH)	177
$CHCl_3$	7.2	Carbon disulfide	193
Benzene	7.3	Acetone (CO)	206
p-Dinitrobenzene	8.4	Cyclopentanone (CO)	217

[a] If H_R is the field at which the TMS resonance is observed and H the field at which the listed resonance is observed, $\delta = (H_R - H)/H_R$. The values of δ will vary with solvent and concentration.

3.1.3 *Measurement of Quadrupole Splittings*

The quadrupole moment of a nucleus may be used as a probe of electric field gradients at their sites. The sample may be a powder, but more information will be available from single crystals. If the isotope is highly abundant, direct observation of the splittings by NMR or pure NQR is possible while for isotopes of low abundance or for cases where few nuclei are experiencing the field gradient to be studied, double resonance techniques have been used quite successfully.

It is often necessary to go to low temperatures to improve the signal-to-noise ratio or increase the spin–lattice relaxation time constant or both. For conductors, various techniques such as electrochemical machining and polishing are often used to increase the surface area and to eliminate surface strains which might mask the desired information.

The spectrum of the sample is recorded using either a field sweep or a frequency sweep and the spacing between the quadrupole split peaks is measured. The experiment can be repeated for various positions of the crystal with respect to any applied magnetic field to obtain crystal structure information or other directional information.

3.1.4 *Measurement of Internal Magnetic Fields*

Paramagnetic, antiferromagnetic, and other magnetically ordered crystals may be studied by NMR. Magnetic fields associated with the paramagnetic ions will cause resonance shifts and splittings from 10^{-4} to 10^{-1} T. These shifts may be orientation dependent, and most information is obtained from single crystals. Low temperatures may be needed to enhance the signal-to-noise but many of the shifts and splittings are themselves temperature dependent. Transient techniques yield the most information, especially multiple-pulse techniques which allow mapping of the magnetic environment of the resonant nuclei.

3.2 Line Widths and Line Shapes

3.2.1 *Line Widths*

a. The line width of a nuclear resonance signal was defined in Section 1.3, Eq. (13). This is the width most often used when the signal is a narrow one such as is seen in a liquid or a gas where the absorption is recorded directly. In solids and in liquids with very broad lines, low-frequency field modulation is commonly used and the recorded signal corresponds to the first derivative of the absorption signal. In this case it is common to measure the distance between the peaks of the first derivative curve. These correspond to the points of maximum slope of the absorption curve. This width is related to the spin–spin relaxation time constant by

$$\Delta f = 1/\sqrt{3}\pi T_2 \tag{27a}$$

or in field units

$$\Delta H = 1/\sqrt{3}\pi\gamma T_2 \tag{27b}$$

for a Lorentzian line. For a Gaussian line, which is more nearly the case for solids,

$$\Delta f = 1/\sqrt{2}\pi T_2 \tag{28a}$$

$$\Delta H = 1/\sqrt{2}\pi\gamma T_2 \tag{28b}$$

It is necessary to ensure that the rf amplitude is small so that no line distortion arises from partial saturation, and that the field (frequency) scan is slow enough to give a faithful reproduction of the signal. If low-frequency field modulation is used, the line may be broadened if the amplitude or frequency of this modulation is too large.

b. In a transient or pulse spectrometer, the spin–spin relaxation time constant T_2 may be measured directly as the time constant for the Bloch decay following a $\pi/2$-pulse. This decay will be exponential if the absorption signal is Lorentzian, and will be Gaussian if the absorption signal is itself Gaussian. In practice, the decay is often some combination of the two, and may show modulation if the resonance line has structure.

For solids the decay may be quite rapid, so large receiver bandwidths are required along with rapid receiver recovery in order that the true transient may be recorded.

3.2.2 *Second Moment Measurement*

a. For a spectrum from a cw spectrometer, the second moment (SM) is usually measured by numerical integration of the first derivative

$$\mathrm{SM} = \tfrac{1}{3}n^2 \frac{\sum_i x_i^3 y_i}{\sum_i x_i y_i} - \tfrac{1}{16}H_{\mathrm{m}}^2 \tag{29}$$

where n is the distance between the measurement points, in field or frequency units, y is the signal amplitude, x is the integer distance from the center of the spectrum, and H_{m} is the peak–peak modulation amplitude expressed in the same units as n.

The second term in the above equation is a correction for sweep modulation broadening. If the frequency of the modulation is a significant part of the line width, this must be corrected too. Again, care must be taken to avoid line distortion due to high radio-frequency power or due to rapid scans.

b. The second moment of a transient signal is the second derivative of the transient decay following a short $\pi/2$-pulse.

$$\mathrm{SM} = (dV^2/dt^2)_{t=0} \tag{30}$$

There will be a slight correction if the receiver bandwidth is too narrow, but of course no correction is necessary for saturation or modulation.

The transient decay for a solid may be very rapid and it may not be possible to get an accurate estimate of the true slope of the transient signal at zero time. In this case, a second pulse can be applied to produce a "solid echo" which forms outside the receiver dead time. The second moment may then be obtained from the equation

$$\mathrm{SM} = (dV^2/dt)_{t=2\tau} \tag{31}$$

where τ is the time between the two pulses. There will be a fourth-order correction to this moment value which can be made small by reducing τ to the smallest value which will allow an echo to be seen.

3.3 Spin–Lattice Relaxation Time Constants

3.3.1 *The Laboratory Frame Time Constant, T_1*

The laboratory frame spin–lattice relaxation time constant may be measured by a progressive saturation technique, making use of Eq. (11). The strength of the radio-frequency magnetic field H_1 must be measured, either by means of an electrical probe or by studying the effect of a known sample.

A more reliable and quicker method is to use transient techniques. The sample can first be prepared in a saturated condition by the application of a $\pi/2$-pulse or by the application of many short pulses spaced closely together. After a waiting time t, the amplitude of the magnetization V_t which has been approaching the equilibrium value V_0 is inspected by a second $\pi/2$-pulse. This amplitude is then fitted to a first-order rate equation to yield T_1

$$V_t = V_0[1 - \exp(-t/T_1)] \tag{32}$$

A second method is to first apply a π-pulse, which inverts the magnetization. The $\pi/2$ inspection pulse will show a negative amplitude for short time t and positive amplitudes for times longer than $T_1 \ln 2$. T_1 may be calculated from the formula

$$V_t = V_0[1 - 2\exp(-t/T_1)] \tag{33}$$

or determined by the time interval $t = T_1 \ln 2$ that gives a zero signal.

The sample may be in any form for these experiments and depending upon the sample material and temperature, T_1's may range from microseconds to weeks. T_1's can be strongly temperature dependent so good temperature control is necessary. Since T_1's are greatly affected by paramagnetic impurities, care in sample preparation is necessary. Liquid samples may need to be outgassed to remove dissolved oxygen which is paramagnetic and will artificially lower the T_1.

3.3.2 *The Rotating Frame Time Constant, $T_{1\rho}$*

In the foregoing section, the spins were relaxed by exchanging energy with the lattice equal to the Zeeman energy, f_0h. The relaxation time constant T_1 is thus a measure of the probability of a spontaneous transition between Zeeman levels which is proportional to the phonon density at frequency f_0. We may inspect the phonon density at different energies by varying the resonance frequency f_0 (field H_0), but for low frequencies sensitivity will be lost. We can overcome this difficulty in two ways.

a. The sample may be adiabatically demagnetized to a low-field value, where relaxation to the lattice is allowed to proceed for a time t. The sample is then remagnetized to the original field where the loss in magnetization is measured with a $\pi/2$-pulse. Assuming negligible relaxation occurs while changing the field, the time constant is calculated from

$$V_t = V_0 \exp(-t/T_{1\rho}) \tag{34}$$

In this way, the signal is generated at the high frequency; so full advantage is taken of the polarization, while the relaxation takes place by exchanging energy with the spins in a low Zeeman energy state. The demagnetization must be slow enough so that the spins may keep their internal equilibrium by means of spin–spin interactions, but fast comparėd to the high-field relaxation time constant.

b. For short relaxation time constants, adiabatic demagnetization in the rotating frame (ADRF), is used. The magnetization is brought into alignment with a strong radio-frequency field H_1 which is then slowly decreased. Relaxation is allowed to proceed for a time t in this reduced field which is then increased to its original value. After rapid shutoff of the radio-frequency power, the transient decay is inspected and Eq. (34) used to calculate the time constant. In this experiment, the sample is always in a high magnetic field H_0, but it "sees" a lower field H_1 and can be thought of as being polarized in this field with an energy level splitting equal to $f_1h = \gamma H_1 h$.

After the inspection pulse in all relaxation time constant measurements, the sample will be in a nonequilibrium state and must be allowed to return to equilibrium by waiting for several time constants in the high field before an experiment can be repeated. Most rotating frame time constant measurements have been performed on solids, but the method is also applicable to viscous liquids.

3.4 Multiple-Pulse Experiments

In multiple-pulse experiments, we use transient techniques to prepare the sample in various states and monitor changes that these states may undergo due to lattice interactions, including the effect of motion and of chemical differences in the nuclear environment.

3.4.1 *Spin Echo Pulse Trains*

If a liquid or mobile solid is placed in a very inhomogeneous magnetic field, the transient decay following a $\pi/2$-pulse may be due entirely to the dephasing of the individual spin vectors brought about by their slightly different resonant frequencies. After complete decay, the spins can be made to refocus, or "echo" by the application of a π-pulse. This echo is caused by the changing of the sign of the individual spin phases by the π-pulse.

After decay of the echo, the π-pulse may be repeated to form a second echo, and this process repeated to form a train of echoes. The echo maxima will decrease in amplitude with a time constant T_2, indicative of the true line width of the sample, not that due to the magnet.

If some of the individual nuclei have moved between the time of the first $\pi/2$-pulse and the echo, or between echoes, they will not be refocused exactly, as their precession frequency will change. The echo amplitudes will thus be reduced and we can use this changed time constant, which will be a function of the time between the pulses, to monitor the degree of movement or diffusion in the sample. The different precession frequencies may be due to chemical shifts, and so we can use this technique to monitor any chemical exchange between different chemical sites. Magnetic field gradients may be introduced which will allow direct measurement of diffusion in the sample. Other pulse sequences can be used to monitor very slow laminar flow in fluids.

3.4.2 *Solid Echo and Line-Narrowing Sequences*

The π-pulses used in the spin echo experiments caused a refocusing of the spin vectors because they had been dephased by a time independent local magnetic field, the field gradient, or the chemical shift. The local field caused by the neighboring dipole moments in a solid fluctuate with time because the spins are undergoing energy level transitions. To achieve refocusing in these time varying fields we must use a $\pi/2$-pulse with a 90° phase shift. The resulting echo is known as a "solid echo" to distinguish it from the more classical spin echo.

A train of properly spaced and phased $\pi/2$-pulses will cause an effective narrowing of the broad resonance lines associated with solids and allows small Knight shifts and chemical shifts to be resolved. The function of the pulse train is to average the $3 \cos^2 \theta - 1$ term in Eq. (15), to zero, thus reducing the broadening caused by the dipole–dipole interaction.

4 Analysis of Materials

In this section we will consider the types of information that NMR can give about a sample. Reference will be made to the previous sections where applicable, and in some cases an example of an application will be given. The literature references cited are a representative sample only, and no attempt has been made to cite all the possible applications. Sample volumes needed and typical analysis times are those for commercially available instruments. Not all of the applications mentioned are possible with present commercial instrumentation, but all are within the scope of present technology.

4.1 Applications to Liquids

In this subsection we will consider applications where the sample is a liquid or must be dissolved in one in order to make a measurement. In some instances the sample may be a mixture of phases, and we will concern ourselves with the analysis of the liquid or "liquid-like" phase.

4.1.1 *Isotope Identification*

a. Nuclear magnetic resonance has a rather limited use as a qualitative detector of isotopes dissolved in liquids owing to the rather low concentrations which are usually encountered. Highest sensitivities are found for ^{1}H and ^{19}F, whereas the heavy metals in general have very low sensitivities. See Section 3.1.1 and Table 1.

b. Quantitative analysis is most often performed on hydrogen. In this case, it is quite common to do quantitative separation of liquid and solid concentration and of viscous and nonviscous liquid components. For example, the amount of oil in corn kernels may be determined rapidly and nondestructively. It is even possible to do this for individual kernels which can then be used for crop breeding.

With commercially available spectrometers, it is generally possible to obtain quantitative accuracies of better than 1% and precisions of 0.1% for samples containing less than 1% of liquid-like hydrogen containing material. Sample volumes of from 2 to 30 ml may be used, the larger volume being necessary for nonhomogeneous bulk samples such as wood chips. Good magnetic field homogeneity will be required to obtain quantitative measurements of narrow line signals from liquid phases in the presence of wider line signals from solid phases.

For samples containing more than 1% liquid-like hydrogen containing material, an analysis is possible in less than 2 min, not counting any sample preparation time. See Section 3.1.1.

For references, see Lee and Anderson (1969), Varian Associates (1966), Conway and Earle (1963), Alexander *et al.* (1964), Sudhakar *et al.* (1971), Zimmerman (1962), and Shoolery and Smithson (1970).

4.1.2 *Compound Identification*

a. Compound identification is undoubtedly the primary use of NMR spectroscopy in liquids. Hydrogen is the nucleus most often observed. In complex compounds, however, ^{13}C spectra are easier to interpret as will be seen in the following example.

The comparison of the hydrogen and ^{13}C spectra of testosterone acetate given in Fig. 11 is helpful in demonstrating the type of information available in the NMR spectra from the two nuclei. The ^{13}C spectrum was ob-

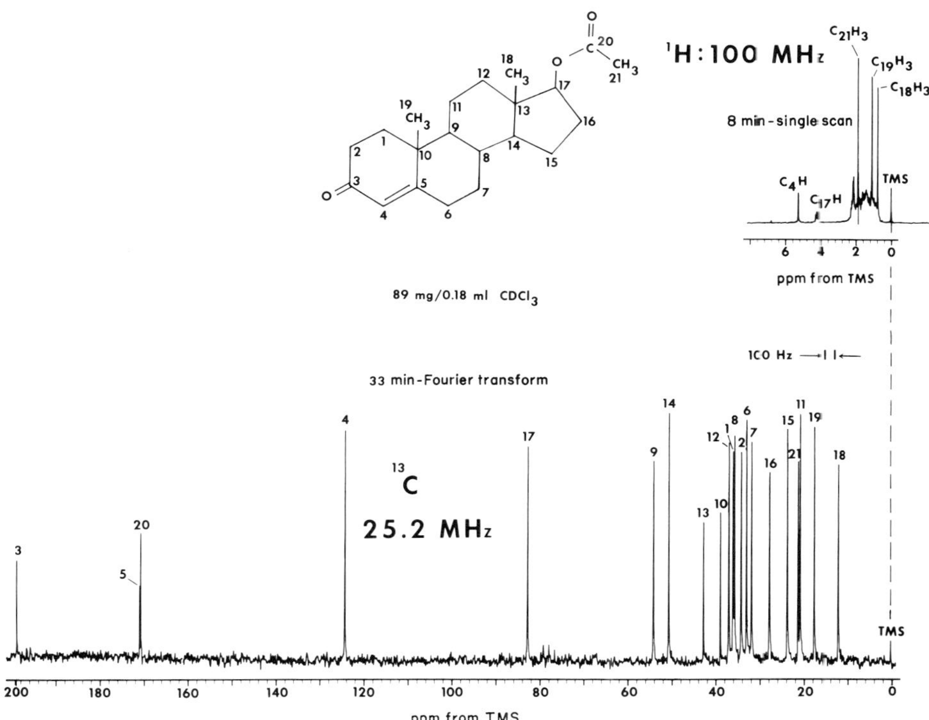

FIG. 11. A comparison of the hydrogen and carbon-13 spectra of testosterone acetate (89 mg in a 0.18-ml solution) shows the much wider spread of the carbon spectrum. The carbon multiplets arising from coupling to the hydrogens have been removed by noise decoupling.

tained using hydrogen noise decoupling and clearly shows the different lines associated with the different carbon sites. Spin multiplets caused by ^{13}C–^{13}C coupling are not easily observed in normal ^{13}C spectra. Because of the low natural abundance of the isotope (1.1%), it is very unlikely that two ^{13}C nuclei will occupy neighboring sites in the same molecule.

The sample will usually be a liquid or soluble solid, signal-to-noise ratios dictating the amount of sample needed. Hydrogen spectrometers commonly use a 5-mm-diam sample tube containing about 0.4 cc of sample, while spectrometers used for heavier nuclei require about 1 cc of sample in an 8- to 15-mm tube. Signal-averaging techniques and Fourier transform spectroscopy are usually needed for low abundant nuclei such as ^{13}C or ^{17}O (see Section 1.8). In this way, good spectra are obtained from less than 1 mg of unknown material.

It is possible to obtain a good hydrogen spectrum in as little as 2 min, while a ^{13}C spectrum will take considerably longer. If selective decoupling is also necessary to aid in interpretation, this will further lengthen the time. See Section 3.1.2.

For reference see Pople *et al.* (1959), Jackman (1959), Varian Associates (1960), Roberts (1961), Varian Associates (1963), Bhacca and Williams (1964), Waugh (1965, 1966, 1968, 1970), Emsley *et al.* (1965), Corio (1966), Hecht (1967), Ionin and Ershov (1970), Bovey (1969), Becker (1969), Farrar and Becker (1971), and Mooney (1968, 1969, 1970).

b. It is possible to obtain quantitative information from high-resolution spectra provided the signal-to-noise ratio is sufficiently high. It is also possible to assay multiple component mixtures with accuracies of better than 5% on samples as small as 25 mg. In this way, the progress of chemical reactions may be followed. (See Nathan y Diaz, 1970.)

4.1.3 *Molecular Motion*

We saw in Sections 1.5.2 and 1.6.1 that the width of a resonance line, or its spin–lattice relaxation time constant T_1, could be used as a measure of the molecular motion in a sample. We can thus use NMR as a viscosity measuring device. One application is in the quantitative determination of oil and water in mixtures such as may be found in natural products, and in the determination of bound and nonbound water in natural products. The oil and the bound water, due to their reduced mobility, have a lower T_1. We can thus separate the two components of the resonance signal of such a mixture (see Sections 3.1.1 and 3.3).

These measurements may be carried out on samples as small as 0.4 ml or as large as 30 ml with accuracies of better than 1%. In simple systems it may be possible to relate the relaxation time constant to some physical parameter of the motion. Recent experiments have been successful in relating the degree of order in biological samples to the presence of malignancies, using samples of a few microliters.

For references see Abragam (1961), Andrew (1958), and Damadian (1971).

4.1.4 *Diffusion*

Diffusion in liquids is best measured using multiple pulse trains. The most accurate results are obtained with time-varying magnetic field gradients, where diffusion constants as low as 10^{-8} cm^2 sec^{-1} have been measured. In principle it should be possible to measure anisotropic diffusion with these methods by applying the field gradient along different axes of the sample. Samples of about 1 ml are commonly used. (See Stejskal and Tanner, 1965.)

4.1.5 *Flow*

Flow in liquids may be measured in several ways using NMR. If lattice relaxation time constants are long, the sample may be passed through a region of alternating magnetic field and the resultant polarization monitored with a tuned pickup coil. Flow rate may be related to the phase change or to the loss in polarization. For slow flow, a multiple-pulse technique may be employed (see Section 3.4.1), which has been used to measure laminar flows as slow as 10^{-2} cm sec^{-1}. Its value in studying flows in biological systems is obvious. Limits on accuracy are dependent upon the degree of turbulence in the flowing system. (See Packer, 1969, and Zhernovoi and Latyshev, 1965.)

4.1.6 *Chemical Exchange*

Chemical exchange can be measured either by studying the line widths and splittings using high-resolution spectroscopy (see Section 3.1.2), or by means of multiple pulse trains (see Section 3.4.1). In the latter case, very fast exchange may be studied, but isotopic substitution may be necessary to prevent masking of information due to nonexchanging resonant nuclei.

A stable high-resolution magnet is required so that the exchanging lines may be resolved. Sample volumes of from 0.4 to 1 ml are usually used which might contain as little as 1 mg of the compound under investigation dissolved in some suitable solvent. Variable temperature control is often required.

For references, see Allerhand and Gutowsky (1964) and Morgan and Strange (1969).

4.2 Applications to Solids

In this subsection we will discuss applications where the sample is a solid or where the presence of a "rigid" lattice is necessary to achieve the desired effects. The technique is nondestructive in all cases in so far as the sample placed into the spectrometer is concerned. Some sample preparation may be necessary, including powdering and annealing, generally to improve the signal-to-noise ratio. If it is not desirable to alter the sample in this way, lowering the temperature or signal averaging may be employed as an alternative.

4.2.1 *Isotope Identification*

As was the case for liquids, NMR is not readily applicable to qualitative isotope identification. In solids the total number of nuclei may be sufficiently high, but if they have high quadrupole moments or interact strongly with near neighbors, the lines may be very broad, reducing the sensitivity. Pure NQR has been used to some extent as an isotope detector using super-regenerative amplifiers. Here again, it has more application in searching for a given isotope.

Commercial NQR spectrometers are available which will search in the 5- to 300-MHz region. About 1 g of sample is commonly required. Conventional NMR spectrometers may be used for quantitative isotope analysis of solids, 1–2 g of sample being required. Accuracies of a few percent are achievable using standard samples for calibration. Analysis time using pulse techniques may be as short as a few seconds.

For references, see Lee and Anderson (1969) and Hahn and Das (1958).

4.2.2 *Compound Identification*

In solids the dipolar interaction tends to mask any small chemical shifts which might be used to distinguish between chemical compounds (see Section 1.5.1). In recent years several experimental procedures have been tried to remove the dipolar broadening and leave the chemical shift information. These techniques, although not readily achievable with present commercial spectrometers, appear to be practicable, especially for ^{19}F and possibly ^{1}H and in systems where the dipolar broadening has been partially removed by molecular motion, in which case complete removal becomes easier.

Recently published double resonance experiments using the quadrupole moment of ^{17}O as a probe of the electronic symmetry of chemical bonds show considerable promise as a means of identifying chemical compounds in solids (see Sections 1.5.5 and 3.1.3).

For references, see Hsieh *et al.* (1972), Waugh and Huber (1967), and Mansfield (1971).

4.2.3 *Molecular Motion*

Molecular motion can affect both of the relaxation time constants T_1 and T_2 by which NMR resonance lines are characterized (see Sections 1.5.2 and 1.6.1). Using cw wideline or pulse spectrometry, the line width and second moment may be measured and related to the rigid lattice crystal structure. Any deviations may be related in simple cases to certain types of motion (see Sections 3.2.1 and 3.2.2). These parameters will only be affected if the motion is of a sufficiently high frequency so that dipolar interactions are averaged.

Measurement of spin–lattice relaxation time constants (see Section 3.3) is perhaps the best means of analyzing molecular motion in solids, particularly using adiabatic demagnetization techniques. In this way, for example, the mean time between atomic jumps in metallic lithium has been measured over the range 10^{-9}–10^{-1} sec.

Sample purity is a prime consideration in performing motion measurements. Paramagnetic impurities will lower T_1 (see Section 1.6.2), and lattice inclusions will promote motion in their vicinity. Variable temperature studies are very important and a good variable temperature system is mandatory. Temperature regulation to better than $\pm 1°C$ is usually required but for work near transition temperatures, regulation to better than 10^{-3}°C may be required. Some sample heating problems may be encountered owing to heat dissipated in the transmitter coil, and for most accurate regulation a total immersion system rather than a gas flow one is preferred.

Samples of about 1- to 2-ml volume are usually used in commercial spectrometers, with somewhat smaller samples being used in noncommercial spectrometers designed to work at temperatures down to and below 2°K.

For references, see Andrew (1958), Abragam (1961), and Ailion and Slichter (1965).

4.2.4 *Crystal Structures*

For simple systems NMR will yield very accurate crystal structure information, owing to the $(\text{distance})^{-3}$ dependence of the dipole and quadrupole interactions (see Sections 1.5.1 and 1.5.5). Both resonance position and second moment information may be used to obtain the necessary data (see Sections 3.1.3 and 3.2.2).

Single crystals will yield the most information, but useful work can also be done on powders, particularly using quadrupole effects. The crystal need not be polished unless it is metallic, when surface preparation may be required to relieve strains and inclusions which will affect the signals. If the spin–lattice relaxation time constant is prohibitively long, introduction of paramagnetic centers by radiation with γ rays or x rays may be helpful. Crystals of about 1 to 2 ml will give good results in most commercial spectrometers but signal averaging and low temperatures may be necessary if low sensitivity or low abundant nuclei or both are being studied.

For references, see Hahn and Das (1958), Abragam (1961), Andrew (1958), and Slichter (1963).

4.2.5 *Lattice Defects*

Lattice defects are best studied using quadrupole effects (see Section 1.5.5). The quadrupole moment of the nucleus is used to probe its electronic environment, which will have low-order symmetry in the vicinity of an impurity, dislocation, or other lattice defect. If the lattice defects are of low abundance, if the nucleus being used as a probe has a low sensitivity, or both, signal-averaging and double resonance techniques may be required (see Section 3.1.3).

Samples of from 1- to 2-ml volume will give satisfactory results in many cases, but very low-temperature studies may have to be used if the sample is a conductor. Spectrometers capable of doing the double resonance experiments in solids are not presently available commercially, but where high sensitivity is not required, commercial machines will give reasonable results.

For references, see Slusher and Hahn (1968), Hartland (1968), Bloembergen (1955), Cohen and Reif (1955), and Lurie and Slichter (1964).

4.2.6 *Magnetic Materials*

Magnetically ordered materials have been widely studied by research workers using NMR. The internal magnetic field at the nucleus due to the electronic structure of the sample can be very high (30–700 T), resulting in NMR frequencies from 40 to 1000 MHz. Resonance positions may be further shifted by quadrupole interactions and Knight shifts (see Sections 1.5.5 and 1.5.6).

The crystallite structure may be studied using multiple-pulse techniques which will map the gradient of resonance shifts in the sample (see Section 3.4.1). Quadrupole splitting rotation patterns from single crystals will give crystal structure details, as will line shapes and second moments (see Sections 3.1.3 and 3.2). The temperature dependence of the resonance position will give information on the temperature dependence of the sublattice magnetization. Single crystals will give the most information, but as most of these materials are conductors, very thin samples will be required to avoid skin-depth problems.

Commercial pulse spectrometers are available which will cover the lower end of the frequency range encountered in these materials, and spectrometers to cover the higher frequencies will no doubt be available when sufficient interest is shown.

See Feldmann *et al.* (1971).

4.2.7 *Metals*

In addition to crystal structure and diffusion data, NMR can yield information on the electronic structure of metals. The Knight shift (see Section 1.5.6) can be related to the net spin imbalance in the s-electron cores due to the spin paramagnetism of the d (or f) electrons. The temperature dependence of the spin–lattice relaxation time constant T_1 can be related to the Fermi distribution of electrons in the metal and it, together with the Knight shift, has proved useful in examining metals.

Signal-to-noise is a problem when working with conductors due to the skin-depth, which limits the penetration of the radio-frequency energy. If we use single crystals we will be looking at only those nuclei near the surface, and the amplitude of the radio-frequency field will not be uniform throughout the sample. Using powders or platelets relieves this problem but we may loose crystal information. The powder may have to be suspended in a liquid to prevent interparticle contact, or sintered in a suitable matrix so that strains introduced during powdering are removed. Thin foils and wires have been used and some directional information is available in these cases.

Variable temperature equipment is a necessity, both for the low temperatures which may be necessary to obtain sensitivity and for the study of the temperature dependence of the conduction electron behavior.

For references, see Abragam (1961), Andrew (1958), Slichter (1963), and Mieher (1966).

4.3 Applications to Gases

Nuclear magnetic resonance measurements have not been widely applied to the study of gases. The major problem is the low signal-to-noise ratio. Usually high pressures must be used to obtain usable signals, although some low-pressure studies have been done using ^{1}H and ^{19}F resonances. Some workers have gone to very high fields to improve sensitivity.

The technique is nondestructive in so far as the sample placed into the spectrometer is concerned. The sample may be a mixture of phases or may be a gas attached to the surface or inside a solid lattice. Most commercial spectrometers are not ideally suited to doing high-pressure studies because they require that the sample and its container be placed inside the sample chamber. The pressure vessel must then be glass or quartz and the sample volume is reduced by the thick walls.

4.3.1 *Isotope Identification*

It is doubtful that NMR will be of much use as a qualitative analysis tool for gases. The sensitivities would in general be too low, causing analysis to be very slow, even for fairly high concentrations of isotopes.

For a specific isotope, NMR could be used for quantitative analysis or for partial pressure measurements. For pressures in excess of atmospheric and for nuclei such as ^{1}H and ^{19}F, good accuracy in a short time should be possible with a precalibrated spectrometer. Measurements on flowing gases using transient techniques would be possible for process analysis. The time spent within the magnetic field would have to be sufficiently high to ensure complete polarization of the sample.

For references, see Abragam (1961) and Andrew (1958).

4.3.2 *Compound Identification*

Compound identification in gases using either cw or Fourier transform techniques will give similar information to that obtainable with liquids. Line widths will be narrow unless the pressure is high, in which case some pressure broadening may occur. Multiple resonance techniques and other instrumental aids to spectral identification applicable to liquid studies can also be applied in the case of gases. Signal averaging will usually be necessary, especially when observing the less sensitive nuclei. If the sample is contained in a quartz pressure vessel, some field distortion due to the vessel may be noticed.

If the gas is absorbed onto a solid matrix, the signal may be a broad line if the molecular mobility is reduced. In this case it may be possible to use multiple-pulse line-narrowing techniques to resolve some chemical shift information (see Section 3.4.2).

4.3.3 *Pressure Measurement*

A measurement of the spin–lattice relaxation time constant T_1 (see Section 3.3) can be used to measure gas pressure. The nuclei in a gas molecule relax via coupling to the fluctuating fields induced by intermolecular collisions. The mechanism is identical to that in a very nonviscous liquid (see Section 1.6.1). In monatomic gases, there are no intramolecular fields, and therefore the only fluctuating fields the nuclei see are those in existence during brief collisions. The T_1's will thus be very long, greater than 10^3 sec in ^{129}Xe at 50 atm.

If sufficient signal-to-noise can be obtained, accurate pressure measurements can be made on gases, the time needed being proportional to T_1. Signal averaging may be necessary for low pressure gases or for dilute systems.

See Abragam (1961).

4.3.4 *Rotational Correlation Times*

As the nuclei in a gas are exchanging energy with the translational and rotational motion, measurement of the exchange rate T_1^{-1} can be used to characterize the quantum states of the motion. Studies of diatomic molecules and simple molecules with more complex symmetry have been undertaken.

These studies are best done using transient techniques. Signal-averaging and variable temperature equipment are almost always necessary. Fourier transformation of the transient spectrum may yield some additional information.

See Abragam (1961).

References

Abragam, A. (1961). "The Principles of Nuclear Magnetism." Oxford Univ. Press (Clarendon), London and New York.

Ailion, D., and Slichter, C. P. (1965). *Phys. Rev.* **137,** A237.

Alexander, D. E., Silvela, L., Collins, F. I., and Rogers, R. C. (1964). *J. Amer. Oil Chem. Soc.* **44,** 555.

Allerhand, A., and Gutowsky, H. S. (1964). *J. Chem. Phys.* **41,** 2115.

Andrew, E. R. (1958). "Nuclear Magnetic Resonance." Cambridge Univ. Press, London and New York.

Becker, E. D. (1969). "High Resolution NMR." Academic Press, New York.

Bhacca, N. S., and Williams, D. H. (1964). "Applications of NMR Spectroscopy in Organic Chemistry." Holdend-Day, San Francisco, California.

Bloembergen, N. (1955). *In* "Defects in Crystalline Solids." pp. 1–32. Physical Soc., London.

Bovey, F. A. (1969). "Nuclear Magnetic Resonance Spectroscopy." Academic Press, New York.

Cohen, M. H., and Reif, F. (1955). *In* "Defects in Crystalline Solids," pp. 44–51. Physical Soc., London.

Conway, T. F., and Earle, F. R. (1963). *J. Office Agr. Chem.* **60,** 256.

Corio, P. L. (1966). "Structure of High Resolution Nuclear Magnetic Resonance Spectra." Academic Press, New York.

Damadian, R. (1971). *Science* **171,** 1151.

Emsley, J. W., Feeney, J., and Sutcliffe, L. H. (1965). "High Resolution N.M.R. Spectroscopy," Vols. I and II. Pergamon Press, Oxford.

Farrar, T. C., and Becker, E. D. (1971). "Pulse and Fourier Transform NMR." Academic Press, New York.

Feldmann, D., Kirchmayr, H. R., Schmolz, A., and Vilicescu, M. (1971). *IEEE Trans. Magn.* **Mag-7,** 61.

Hahn, E. L., and Das, T. P. (1958). "Nuclear Quadrupole Spectroscopy." Academic Press, New York.

Hartland, A. (1968). *Proc. Roy. Soc.* A**304,** 361.

Hecht, H. G. (1967). "Magnetic Resonance Spectroscopy." Wiley, New York.

Hsieh, Y., Koo, J. C., and Hahn, E. L. (1972). *Chem. Phys. Lett.* **13,** 563.

Ionin, B. I., and Ershov, B. A. (1970). "N.M.R. Spectroscopy in Organic Chemistry." Plenum Press, New York.

Jackman, L. M. (1959). "Applications of N.M.R. Spectroscopy in Organic Chemistry." Pergamon Press, Oxford.

Lee, K., and Anderson, W. A. (1969). "A Table of Nuclear Spins, Moments and Magnetic Resonance Frequencies." Varian Associates, Palo Alto, California.

Lurie, F. M., and Slichter, C. P. (1964). *Phys. Rev.* **133,** A1108.

Mansfield, P. (1971). *J. Phys. C* **4,** 1444.

Mieher, R. L. (1966). *In* "Semiconductors and Semimetals" (R. K. Willardson and A. C. Beer, eds.), Vol. 2, pp. 141–187. Academic Press, New York.

Mooney, E. F. (1968, 1969, 1970). "Annual Review of N.M.R. Spectroscopy," Vol. I–III. Academic Press, London.

Morgan, R. E., and Strange, J. H. (1969). *Mol. Phys.* **17,** 397.

Nathan y Diaz, J. (1970). "Introduccion a la Resonancia Magnetica." Limusa-Wiley, Mexico.

Packer, K. J. (1969). *Mol. Phys.* **17**, 355.
Pople, J. A., Schneider, W. G., and Bernstein, H. J. (1959). "High Resolution Nuclear Magnetic Resonance." McGraw-Hill, New York.
Roberts, J. D. (1961). "An Introduction to the Analysis of Spin-Spin Splitting in High Resolution N.M.R. Spectra." Benjamin, New York.
Shoolery, J. H., and Smithson, L. H. (1970). *J. Amer. Oil Chem. Soc.* **47**, 153.
Slichter, J. C. (1963). "Principles of Magnetic Resonance." Harper, New York.
Slusher, R. E., and Hahn, E. L. (1968). *Phys. Rev.* **166**, 332.
Stejskal, E. O., and Tanner, J. E. (1965). *J. Chem. Phys.* **42**, 287.
Sudhakar, S., Steinberg, M. P., and Nelson, A. I. (1971). *J. Amer. Oil Chem. Soc.* **48**, 11.
Varian Associates Staff (1960). "N.M.R. and E.P.R. Spectroscopy." Pergamon Press, Oxford.
Varian Associates Staff (1963). "High Resolution N.M.R. Spectroscopy Catalog." Varian Associates, Palo Alto, California.
Varian Associates Staff (1966). Varian PA-7 Process Analyzer: N.M.R. Analyzer for Industry. Bull. INS 1469. Varian Associates, Palo Alto, California.
Waugh, J. S. (ed.) (1965, 1966, 1968, 1970). "Advances in Magnetic Resonance," Vols. I–IV. Academic Press, New York.
Waugh, J. S., and Huber, L. M. (1967). *J. Chem. Phys.* **47**, 1862.
Zhernovoi, A. I., and Latyshev, G. D. (1965). "Nuclear Magnetic Resoance in a Flowing Liquid." Consultants Bureau, New York.
Zimmerman, J. R. (1962). *In* "Methods of Experimental Physics" (D. Williams, ed.), Vol. 3, pp. 359–440. Academic Press, New York.

CHAPTER 14

Raman Spectrometry

J. E. Katon

Miami University
Oxford, Ohio

Introduction

Although not a new technique for materials analysis, Raman spectrometry has become a much more powerful and convenient tool in the last few years due to the advent of the laser source. The method gives data which are similar to those obtained from infrared spectroscopy, but offers a number of advantages over infrared with many samples.

The theoretical principles are briefly outlined and the attendant experimental advantages of Raman spectrometry over infrared spectroscopy are discussed. The widest application at present is for qualitative analysis and the advantages and limitations of Raman spectrometry in this area are given. Sample handling considerations are discussed and a brief discussion of quantitative analytical potential is given.

It is pointed out that Raman spectrometry has several distinct advantages over other methods for the qualitative analysis of corrosive liquids, highly hydrogen bonded systems, aqueous solutions, polymers, and inorganic materials such as glasses, minerals and fused salts.

1 Theoretical Principles

In the 1930s, material analysis by vibrational spectroscopy was practically synonymous with analysis by Raman spectrometry. The experimental difficulties were severe, however, as they were with most instrumental analytical techniques by today's standards. The development of reliable double-beam infrared absorption spectrophotometers in the middle and and late 1940s led to the replacement of Raman spectrometry by infrared absorption spectroscopy as the method of choice for analytical applications utilizing vibrational spectroscopy. In the last few years, there has been a great resurgence in Raman spectrometry, due primarily to the development of the laser source, and, more recently, the advent of relatively low cost Raman spectrometers. It seems likely that it will, in the next few years, become a standard complement to infrared spectrophotometry in most laboratories for analysis of materials through their vibrational spectra. Throughout this chapter we will concern ourselves only with vibrational spectra since the vast majority of analytical applications of Raman spectrometry are concerned with such spectra. Although both electronic and rotational Raman spectra may be obtained, and in certain cases may be very useful for analytical work, their use is limited to quite narrow areas.

It has, by now, become conventional to compare Raman spectrometry with infrared absorption spectroscopy in any discussion of the analytical usefulness of the former. The type of data obtained is the same in both methods, namely, the difference in vibrational energy levels of molecules. The experimental techniques are quite different, however, as are the details of the information obtained. Certainly this is to be expected in view of the very different physical bases of the two processes. Whereas infrared absorption spectroscopy is an example of a very common spectroscopic phenomenon, resonance absorption, Raman spectrometry depends on an inelastic scattering process.

In Fig. 1 the difference in the two processes is indicated from the point of view of a simple energy level diagram, which represents three molecular vibrational energy levels in the ground electronic state. The three levels are characterized by values of the vibrational quantum numbers 0, 1, and 2, respectively. An infrared photon of the proper frequency will lead to the transition A, resulting in infrared absorption. If the molecule is exposed to radiation in the visible region of the spectrum, however, and the incident photon's energy is insufficient to lead to an electronic transition, the scattering process may be observed. One can view this (although the view is not strictly correct from a theoretical basis), as involving a "sticky" collision of a photon with a molecule. This collision, represented by B, in-

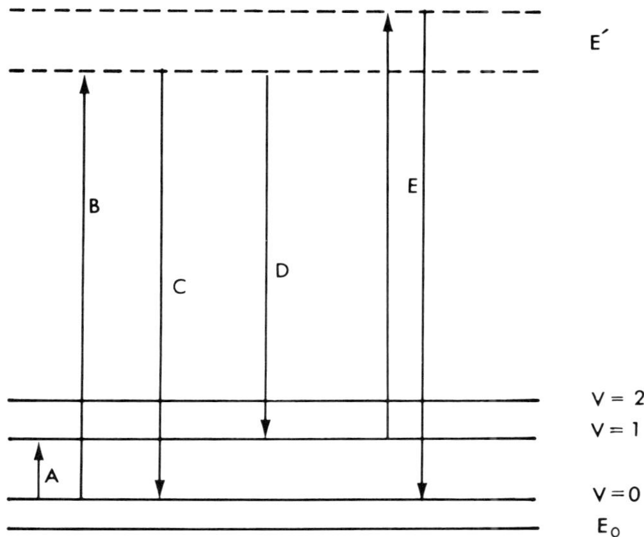

FIG. 1. Simplified energy level diagram for a hypothetical molecule showing transitions corresponding to molecular absorption and scattering processes.

creases the energy of the molecule to some very temporary value represented by E'. As the photon then rebounds from the molecule three possible situations may arise.

1. The scattered photon may possess the same energy as it had originally; that is, the collision may be elastic. This corresponds to the transition C in Fig. 1 and gives rise to what is commonly called Rayleigh scattering. This is the most common occurrence.

2. The scattered photon may possess less energy than it originally had. This process is represented by transition D. It corresponds to an inelastic collision in which the molecule gains energy from the photon. The quantum restrictions of the molecule's energy levels demand, however, that it acquire only the amount of energy corresponding to the difference between two vibrational energy levels. It is readily seen from the diagram that the difference in energy (frequency) of the incident photon and the scattered photon is exactly the same frequency as that of the infrared photon absorbed during the corresponding transition from the lower level to the higher level. This process is called Raman scattering.

3. If the molecule is initially in an excited vibrational level due to thermal excitation, the collision may be inelastic in the sense that the photon gains energy from the molecule (E). In this case the difference in energy

of the scattered photon and the incident photon is again required by the molecule's quantum restrictions to be equivalent to the energy separation of the two vibrational levels.

These last two processes typically occur to about 0.1% of process 1. Clearly, process 2 will occur considerably more often than process 3 for most vibrations at normal temperatures since the ground state will normally be much more highly populated than will the excited state.

If one irradiates a polyatomic molecule with monochromatic light of frequency $\bar{\nu}_0$ cm^{-1} and passes the scattered radiation through a spectrograph, one observes the intense Rayleigh line, unshifted at $\bar{\nu}_0$, and in addition, a series of lines at $\bar{\nu}_0 + \bar{\nu}_1, \bar{\nu}_0 + \bar{\nu}_2, \bar{\nu}_0 + \bar{\nu}_3, \ldots$ cm^{-1} and a second series at $\bar{\nu}_0 - \bar{\nu}_1, \bar{\nu}_0 - \bar{\nu}_2, \bar{\nu}_0 - \bar{\nu}_3, \ldots$ cm^{-1}, $\bar{\nu}_1, \bar{\nu}_2, \bar{\nu}_3, \ldots$ being the frequencies corresponding to various vibrational excitations of the molecule (fundamental vibrational modes of the molecule, e.g., CH stretch, CC stretch, CH bend, etc.). The second series is the most intense as mentioned earlier. It can be readily seen that the data of interest are given by the values of $\Delta\bar{\nu}$, or the differences between the frequencies of the measured lines and the frequency of the Rayleigh line, and not the absolute frequencies of the lines. Only the former are characteristic of the molecule, the latter being also a function of the frequency of the incident radiation. It is for this reason that Raman data are given in terms of Raman *shifts*, $\Delta\bar{\nu}$ cm^{-1}.

Although the simple energy level diagram in Fig. 1 implies that the data obtained from Raman spectrometry and infrared spectrophotometry are the same, closer observation of the details of the two methods reveals significant differences. Infrared absorption requires a change in the molecular dipole moment, whereas Raman scattering requires a change in the molecular polarizability. Furthermore, the intensity of the absorption or scattering is related to the magnitude of the changes in these two molecular properties during the course of a vibrational excitation. The implications of these differences to the observed spectra are extensive.

If the molecules of the material being studied possess some significant amount of symmetry, only certain of their vibrations will result in a change in dipole moment. As a result only a portion of their total vibrations will give rise to infrared absorption. Although the same sort of statement can be made about their Raman spectra, i.e., only certain vibrations will result in a change in polarizability and therefore an observed Raman band, in nearly all cases more fundamental vibrations will be Raman active than will be infrared active. Furthermore, nonfundamental vibrations (overtones, combinations, etc.) tend to be much more weakly allowed in the Raman spectrum than in the infrared spectrum, and, in fact, are only rarely observed. These two features lead to a rather significant conclusion: In general, the Raman spectrum possesses more information concerning a symmetrical molecule's fundamental vibrational spectrum than the infrared spectrum while at the same time it is less cluttered with extraneous features which may be misleading in data interpretation. Some care must be utilized in the interpretation of this statement since nonfundamental vibrations are sometimes useful in identifying materials. This results in the common observation that both methods should be used if they are available. Nevertheless, it serves as an indication of the potential power of Raman spectroscopy.

If the molecules of the material in question possess little or no symmetry, the two methods remain complementary because of the wide disparity often noted in the intensities of fundamental vibrations that are both infrared and Raman active. In general, vibrations of the molecule which involve polar bonds are strong in the infrared and those involving mostly nonpolar bonds are strong in the Raman. The very weakly allowed characteristic of nonfundamental vibrations in the Raman spectrum persists with nonsymmetric molecules so that again the Raman spectrum is generally simpler than the infrared spectrum.

There is one other characteristic of Raman spectrometry which is distinctive and potentially useful. If the incident light utilized in exciting a Raman spectrum is plane polarized, one finds that the scattered Raman light related to certain of the molecular vibrations retains some or all of the polarization properties of the incident light. Experimentally, this effect is measured by placing a polarizer just prior to the entrance slit of the monochromator and measuring the intensity of the scattered light when this polarizer is aligned both parallel to and perpendicular to the plane of polarization of the incident light. The results are expressed in terms of the depolarization ratio ρ of the Raman band being investigated, where $\rho = I_{\perp}/I_{||}$ and $I_{\perp}$ is the intensity of light measured when the polarizer is perpendicular to the plane of polarization of the incident light and $I_{||}$ its intensity when the polarizer is parallel to the plane of polarization of the incident light. A detailed theoretical treatment of the effect shows that ρ can vary from zero to 6/7 for isotropic samples and, in general, has lower values for the more symmetrical vibrations of the molecule. A detailed treatment of polarization effects is beyond the scope of this discussion but can be found in several of the treatments of Raman spectrometry listed in the bibliography. Not much work has been published in which values of ρ have been correlated with structural features of molecules, but it seems to be an area worthy of further investigation.

From the foregoing discussion it is apparent that the laser is an ideal source for Raman spectrometry for three major reasons.

1. The radiation emitted from the laser is quite highly monochromatic.

This is a necessity for good quality Raman spectra since it leads to only one set of Raman shifts in the observed data.

2. The radiation is very intense. This factor is also quite important because of the inherent weakness of the Raman effect.

3. The radiation is polarized, making depolarization ratio measurements much more convenient.

A further characteristic of Raman scattering which is involved in reason 2 and which affects the choice of laser is the dependence of Raman intensity on the wavelength of the incident radiation. This dependence is governed by the fundamental relationship, $l \propto 1/\lambda^4$, where l is the intensity of scattered radiation and λ is the wavelength of the incident radiation. Clearly, incident radiation in the blue region of the visible spectrum will result in a considerably more intense Raman spectrum than will radiation in the red region. (As an aside, it might be mentioned that the conventional detectors of Raman radiation are photomultiplier tubes. These are more sensitive in the blue region and this gives an added advantage to using short wavelength light as a source.) With a good many organic molecules, however, blue light leads to fluorescence of the sample. This gives a strong background which tends to obscure the Raman spectra and so it is often advantageous to use red light, where the high power of the laser is extremely advantageous. The most common exciting lasers in use today are the argon ion laser (488.0 nm or 514.5 nm) and the helium-neon laser (632.8 nm). In the majority of cases, the argon ion laser is preferable for the production of high-quality spectra. A block diagram of a laser Raman spectrometer is given in Fig. 2.

Of course, the fact that Raman spectrometry deals experimentally with radiation in the visible region of the electromagnetic spectrum leads to experimental details quite different than those encountered in the infrared region. In general, experimentation is easier in Raman spectrometry than in infrared spectrophotometry. Visible-type optics are used throughout and, to the analyst, the most significant feature of this is the use of glass sample cells rather than the much more difficult to handle alkali halides and polyethylene used in the infrared. Because of the shorter *wavelength* region covered, a Raman spectrometer will give data in the region 4000–50 cm^{-1} with one grating and no auxiliary filters (the grating must be of considerably better quality than infrared gratings, however). Coverage of this range in the infrared spectrum normally requires six or more gratings along with their associated filters, two sources, and two detectors. In addition, a reasonably high-quality infrared spectrum over this range would require at least three different path-length cells, at least two of which were constructed of different optical materials.

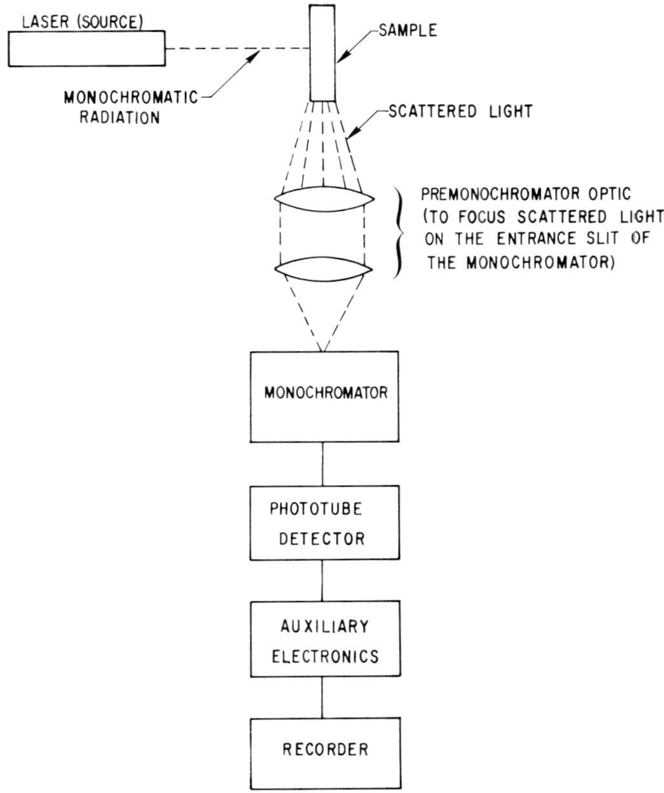

FIG. 2. Block diagram of a laser Raman spectrometer showing essential features.

The major disadvantages of Raman spectrometry include the high stray light level (nearly all of this arises from the much more strongly scattered Rayleigh line), which requires multiple monochromators for removal; and much more precise mechanical tolerances in the instrument. These disadvantages are, in general, taken care of by the instrument manufacturer. Of more interest to the analyst are problems of fluorescence and absorption, which may make it difficult to obtain good spectra with some samples (probably no more than 10%), and the difficulty of obtaining good intensity measurements in Raman spectrometry. Raman intensities are linear with concentration but are also functions of other parameters, some of which are instrumental. It is therefore more difficult to make precise intensity measurements in a Raman spectrum than in the corresponding infrared spectrum.

2 Applications

2.1 Qualitative Analysis

At this time the major analytical application of Raman spectrometry is for qualitative analysis. In principle, the qualitative analytical aspects of Raman spectrometry follow those of infrared spectrophotometry rather closely. This is especially true if one is using the method of comparison of an unknown spectrum with reference spectra run under the same, or similar, conditions. In this method, of course, one simply compares spectra with respect to observed frequencies and relative intensities of spectral features until a match is found, thereby identifying the unknown. Raman spectrometry suffers in comparison to infrared when using this method because of the present relative paucity of spectral files or libraries of known compounds. This will probably be a short-lived disadvantage, however, as Raman spectrometry becomes more widely used. On the other hand, it has an advantage in that a wider spectral range is covered. For a number of years this led to the fact that only Raman spectrometry was good for the identification of most inorganic compounds since most of their vibrational frequencies were lower than the low-frequency cutoff of normal infrared spectrometers. Much of this advantage of Raman spectrometry for inorganic compounds was obviated in the 1960s with the advent of extended range, grating infrared spectrophotometers, but the region below 200 cm^{-1} is still much more easily studied by Raman techniques than by infrared.

In the area of qualitative analysis through use of the group frequency concept, the complementary principle of Raman and infrared techniques becomes very important. A number of significant functional groups are considerably more evident in the Raman spectra of compounds containing them, than in the corresponding infrared spectra. On the other hand, the infrared absorption technique is superior for others. As was implied in Section I, those functional groups which are polar tend to give good infrared group frequencies; those which are nonpolar tend to give Raman group frequencies. The term "good" is, in this context, used to indicate relatively intense and easily identifiable bands. The "good" characteristic of group frequencies relating to the range of the spectral region over which the band occurs in different compounds is related to the theoretical treatment of molecular vibrations and is beyond the scope of this treatment. It might be noted here that any good correlation chart for infrared spectra can be used for Raman spectra as well. The relative intensities are different, but the frequency ranges are identical.

In addition, groups or atoms which contain a large number of electrons tend to give large changes in polarizability during vibrations and thus lead to strong Raman bands. Such groups may or may not lead to strong infrared absorption bands.

At present, it is difficult to compare infrared and Raman usefulness for many functional groups since the Raman group frequency concept has not been nearly as fully explored as has that of infrared. Several research groups are investigating the general area of Raman group frequencies, however, and much more should be known in a few years.

At present, however, it is possible to make several useful generalizations:

1. Functional groups or structures which involve homonuclear bonds generally give stronger bond stretching Raman bands than infrared. Examples are C—C, S—S, N—N, and the metal-metal bonds such as Pb—Pb which are currently of great interest in inorganic chemistry. If the groups are multiply bonded, C═C, C≡C, N═N, etc., they tend to be even stronger in the Raman spectrum because of their higher polarizability. Figure 3 reproduces the Raman spectrum of propynoic acid and illustrates the strength of the C≡C stretching mode (2129 cm^{-1}). It is often impossible to detect the carbon-carbon triple bond stretching mode in complex organic molecules or the carbon-carbon double bond stretching mode in trans disubstituted olefins by infrared spectrometry. However, these vibrational modes are invariably very strong in the Raman spectrum. The S—S linkage is a classic example of a group that gives strong Raman bands and weak infrared bands, making it very difficult to detect in the infrared but readily apparent in the Raman spectrum.

2. Molecules containing heavy atoms generally have much stronger Raman bands than infrared bands associated with motions involving the heavy atom bonds. This is due to the high polarizability of these bonds arising from the large number of electrons in their vicinity. Most organometallic compounds show this behavior as well as the organometalloids containing elements such as arsenic, selenium, germanium, etc., and organic iodides and bromides.

3. Hydrogen bonding does not intensify the Raman spectrum as it does the infrared spectrum. As a result, the broad, strong bands observed in the infrared spectrum of hydrogen bonded compounds are weak or nearly absent in the Raman spectrum. Although this makes Raman spectrometry a poor method for detecting hydrogen bonding, it makes it a very good method for observation of secondary features in the structures of highly hydrogen bonded materials such as sugars and their derivatives. Such compounds yield Raman spectra with sharp, distinct bands due to motions of the skeleton and other nonhydrogen bonded groups. These features are often obscured in the infrared spectrum by the very broad, ill-defined absorption bands due to the hydrogen bonded groups.

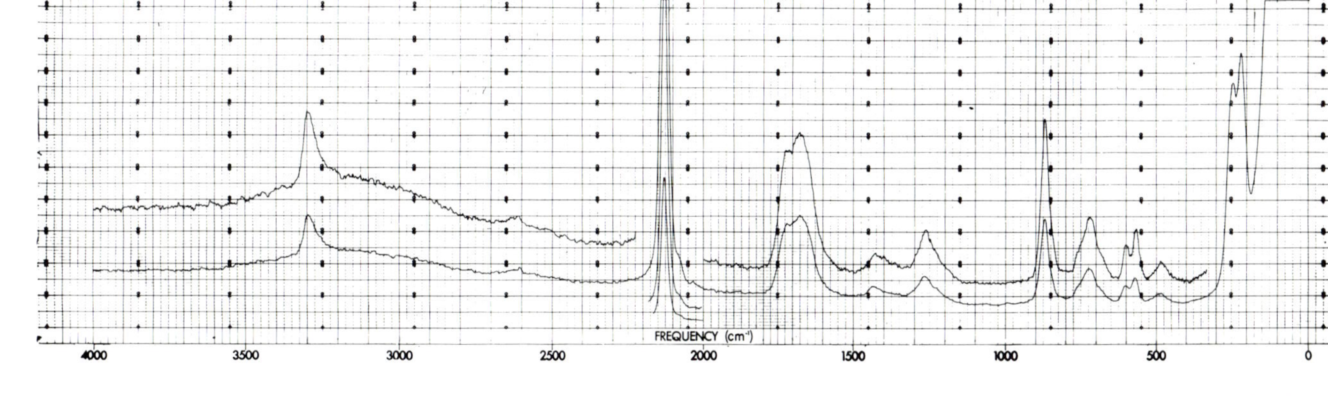

Fig. 3. The Raman spectrum of propynoic acid. Note the very large intensity of the C≡C stretching vibration relative to all other vibrational modes. [From Katon and McDevitt, 1965.]

4. Local symmetry effects lead to intensity reversals in the Raman and infrared spectra which are often important for structural conclusions or for confirmatory evidence concerning the presence of a functional group. Thus, there are two stretching modes in the groups NO_2, CO_2^-, COC, and CH_2. The antisymmetric stretching mode, that is, the one in which one bond is being elongated while the other is simultaneously being compressed, is stronger in the infrared spectrum and is usually at higher frequency. The symmetric stretching mode, the one in which both bonds elongate or compress simultaneously, is stronger in the Raman spectrum at a somewhat lower frequency. An interesting example of this effect in the determination of a structural feature is found in the determination of the hydrogen bonded structure of organic carboxylic acids. These compounds are known to exhibit both the hydrogen bonded dimer structure and the hydrogen bonded polymer structure in condensed phases. The hydrogen bonded dimer has sufficient local symmetry in the hydrogen bond ring to allow one to describe the two carbonyl stretching vibrations as symmetric and antisymmetric. The dimer structure exhibits a strong carbonyl stretching mode at about 1720 cm^{-1} in the infrared (antisymmetric) and a strong Raman carbonyl stretching mode at about 1680 cm^{-1} (symmetric). The polymer structure exhibits carbonyl stretching modes in the infrared and Raman spectra that coincide in frequency however (1700-1740 cm^{-1}). Examples of these effects are furnished by the infrared and Raman spectra of iodoacetic acid (Figs. 4 and 5). Note the intensity reversal of the two CH_2 stretching modes in the Raman and infrared as well as the difference in carbonyl stretching frequencies.

5. The effects of a large number of electrons in a group are extended to hydrogen atoms attached to the group. Thus, olefinic and acetylenic carbon–hydrogen stretching modes are stronger in the Raman spectrum than in the infrared spectrum. This is also true of sulfur–hydrogen stretching modes.

6. Aromatic compounds give strong Raman spectra. In particular, they always give a very strong band associated with the symmetrical expansion and contraction of the ring. (Even saturated ring systems possess this characteristic Raman spectral feature.) For instance, substituted benzenes give a characteristic, very strong Raman band near 1000 cm^{-1}. Some preliminary evidence exists that Raman spectrometry may be much more valuable than infrared for identifying and characterizing heterocyclic ring systems by the frequency of this so-called "ring breathing" mode, since it is so strong in nonaromatic ring systems, as well. Figure 6 illustrates this band in aromatic ring systems as it occurs in the Raman spectrum of diphenyl ether.

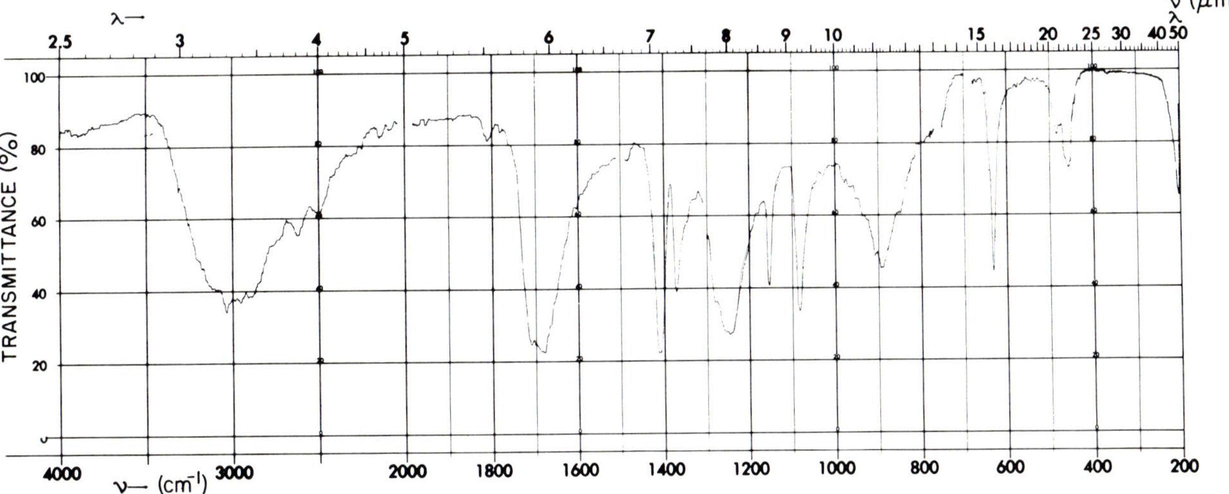

FIG. 4. The infrared spectrum of a mixed mull of iodoacetic acid. The CH_2 stretching modes are barely visible at 3058 and 2985 cm^{-1} superimposed on the very strong OH stretch. The carbonyl stretch is unusually low in this compound being found at 1680 cm^{-1}. [From Katon and Carll, 1971.]

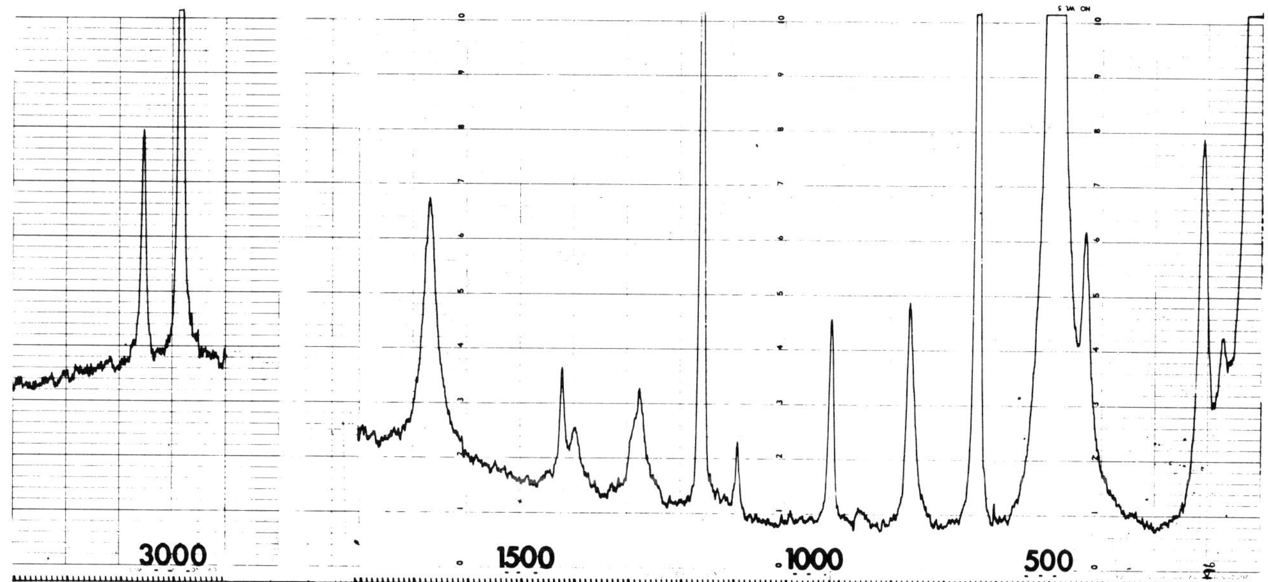

FIG. 5. The Raman spectrum of crystalline iodoacetic acid. The carbonyl stretch is at 1669 cm^{-1}. [From Katon and Carll, 1971.]

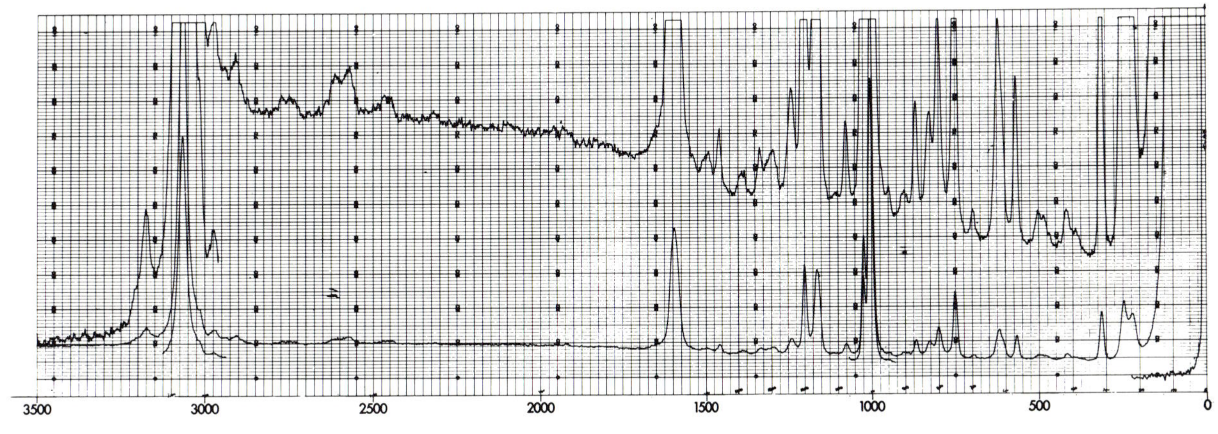

FIG. 6. The Raman spectrum of diphenyl ether. Note the very strong "ring-breathing" band just above 1000 cm^{-1}. [From Katon *et al.*, 1964.]

The sensitivity of Raman spectrometry is a very strong function of the compound being sought, just as it is in the infrared. Clearly, the stronger the Raman band that is possessed by a substance, the smaller the amount of that material that will be detectable. The technique has been applied to the detection of various hydrocarbons in mixtures, but it is not generally used for low levels of compounds. As a general estimate, one is not likely to detect a component present in concentration of less than about 5%, but this figure can vary rather widely. Probably the most potential for detection of low concentrations of materials by Raman spectrometry is found in the study of aqueous solutions, since these solutions do not lend themselves readily to study by other techniques.

The time required for the qualitative analysis of a sample compares favorably with that for infrared analysis. Sample preparation time is generally less in the Raman, and assuming that no problems with absorption or fluorescence are encountered, a spectrum can be obtained in about 30 min or less. Interpretation of the spectrum requires normally about 30 min but varies widely, depending on the compound.

2.2 Sample Considerations

With the advent of the laser source and some research on sampling procedures that has followed it, many of the former sample limitations of Raman spectrometry have been removed. It is likely that at least some of the present difficulties will be reduced by further development of sampling techniques as time goes on. At present, the major limitations are sample fluorescence and sample absorption of the exciting radiation.

If the material whose spectrum is to be obtained fluoresces, one has available a few approaches to solve the difficulty. The easiest is probably to change the wavelength of the exciting radiation. This often produces striking results. For instance, terphenyl shows a very strong fluorescence when excited by the 488.0-nm argon ion line, but gives a good Raman spectrum when excited by the 514.5-nm line. More usually one must go to a red exciting line such as the helium–neon laser line at 632.8 nm if the sample fluoresces. New lasers are being developed which have four or five widely separated lines of sufficient power to excite a good Raman spectrum. Further development along these lines promises to reduce the cost of having several available excitation lines which presently requires at least two lasers. A second empirical approach involves "baking" the sample in the laser beam. The process or processes occurring under this condition are not understood at present, but nevertheless sample fluorescence is often greatly reduced after a period of irradiation with the laser beam. The time of irradiation necessary varies widely. With many compounds 5–10 min are sufficient, but with others, particularly large biologically interesting molecules, 12 or more hours are required.

If the sample contains a fluorescent impurity one can often remove it, or reduce its effects, by sample treatment providing the sample is a liquid or can be dissolved in a solvent. The techniques involved here were developed a number of years ago by Raman spectroscopists. Careful distillation in a microdistillation apparatus, treatment with decolorizing charcoal, or filtering through Millipore* filters often helps greatly. One technique developed by practicing Raman spectroscopists previous to the laser sources but which appears to be not widely known outside of the field is the use of fluorescence quenchers. If the material is in the liquid form, a drop or two of nitrobenzene or nitrotoluene often greatly reduces the sample fluorescence. Sodium iodide is often useful in this regard if the sample is an aqueous solution.

As with fluorescence, sample absorption can sometimes be circumvented by changing the exciting wavelength. In other cases, satisfactory spectra may sometimes be obtained by "back scattering." In this approach, the sample is illuminated with the laser beam at an angle and the light which is scattered at a "reflection" angle is passed into the monochromator. This technique has been used to obtain excellent Raman spectra of essentially opaque materials. One must recognize that spectra so obtained are from surfaces, however, and may not always be representative of the bulk sample. In addition, care must be used with such materials since thermal and photodecomposition can occur if the laser beam is too intense. Recently these problems have received a good bit of attention by research workers and the thermal or photodecomposition problem or both have been largely overcome by the utilization of a rotating sample. The laser beam excites a small area of the sample in any normal experimental procedure (excluding those involving multipass cells) with a rather large energy density. If the sample absorbs this radiation it quickly becomes very hot. If the area of the sample being irradiated is continually changed by rotation of the sample, however, local thermal effects appear to be nearly completely eliminated. There is no attendant loss in quality of the Raman spectrum, and there may be, in fact, an increase in quality due to the resonance Raman effect. This effect produces much stronger Raman scattering when the incident radiation is very close in frequency to that required for an electronic transition in the material being studied.

Apparatus, techniques, and results obtained utilizing a rotating sample have been described by Kiefer and Bernstein (1971). It should be noted

* Trademark of Millipore Corporation, Bedford, Massachusetts.

that this procedure does not affect fluorescence, however, and one must attempt to solve fluorescence problems by the approaches mentioned earlier.

If absorption or fluorescence is not too troublesome, sampling procedures for Raman spectrometry are, in general, as easy or easier than those in the infrared. Vapors and liquids are easily handled even in quite small quantities. Good spectra of liquids in quantities of the order of 1.0 μg or less can be obtained. This makes Raman spectrometry useful for obtaining spectra of gas chromatographic fractions, particularly since it is much easier to collect such fractions in glass capillaries which can be directly used for recording spectra. Infrared techniques require sample transfer and beam condensing optics, which require more time, in general. Powders are readily handled in small quantities (1.0 μg or less), without further preparation such as pressing into pellets. Many modifications of sampling techniques are described by Gilson and Hendra (1970). The differences involve only details, however, and are primarily of interest for microsampling.

Raman spectrometry has a distinct advantage for studying certain types of samples. Single crystals and single fibers are readily studied by Raman spectrometry, but are much more difficult to study by infrared. In addition, the Raman spectra of macroscopic solids, such as manufactured articles, can often be obtained directly, whereas infrared spectra would require lengthy sample preparation. Raman spectra of screwdriver handles and gear wheels have been published to illustrate this advantage.

2.3 Quantitative Analysis

Although Raman spectrometry was used in quantitative analysis by several industrial laboratories in the 1930s and 1940s, infrared replaced it soon thereafter. Neither are so widely used at present because of the much better sensitivity of newer methods such as gas-liquid chromatography. Nevertheless, a few comments are in order since in certain samples Raman spectrometry may be the method of choice due to the nature of the sample or the requirements of sampling conditions.

Raman spectrometry is essentially a single-beam spectroscopic method and as such, it suffers from all of the disadvantages of this method. Any change in experimental conditions during an analysis will cause problems. Thus, source intensity, sample temperature, and other instrumental effects which are not held constant will affect the results. In addition, sample positioning must be carefully controlled. If one is to measure the intensity of Raman bands in various samples one must be certain that the samples are irradiated at a constant flux with identical geometries. Furthermore, the light is scattered in all directions and the proportion entering the instrument must remain constant. The fraction of the irradiating light which excites the sample and the fraction scattered which enters the monochromator is a function of the refractive index of the sample and the sample cell. Although this is also true for infrared methods, they are not such strong functions of refractive index. It is therefore necessary, if good results are to be obtained, to prepare a calibration curve with a set of standards that are recorded in an identical manner. When this is done, an accuracy of 1% or better can be obtained.

The method is not very good for minor components because of the relative weakness of Raman scattering. It has certain advantages in mixtures, however, because the relative simplicity of the Raman spectrum makes it easier to find isolated bands due to a single component. In addition, the intensity of a Raman band is linearly related to concentration of the species causing the band. This somewhat simplifies the associated calculations.

The time required for a quantitative analysis is short if a calibration curve is available. Certainly no more than 5 min is required for recording and analyzing the data. To this must be added sample preparation time and especially time to prepare the calibration data. The time required for calibration probably prevents the use of the method in many cases unless very many similar samples are to be analyzed.

2.4 Specific Applications

Recent monographs on Raman spectrometry (Gilson and Hendra, 1970; Tobin, 1971) have listed a number of applications involving specific compounds or classes of compounds. The literature has been growing rapidly and it is not difficult to find many papers in which Raman spectrometry has allowed investigators to obtain information which is impossible, or at least very difficult to obtain by any other means. As a result, we will limit our attention here to somewhat general applications where Raman spectrometry possesses distinct advantages over other methods.

2.4.1 *Corrosive Liquids*

The advantage of Raman spectrometry in studying corrosive liquids is primarily an experimental one. Spectroscopic methods are perhaps the only really good way of studying liquid structures of pure materials and especially the structures of solutes in a solution. Infrared spectroscopy has been widely used to identify and detect various species in solution. It is nearly impossible to find materials which are transparent to infrared radiation which will resist attack by strong inorganic acids and bases and other very corrosive liquids, however. The use of glass, pyrex, or quartz cells in Raman spectrometry provides a suitable containment material which is highly inert. Thus, sulfuric acid solutions or strongly alkaline solutions are readily studied by Raman spectrometry. Since a number of highly reactive species are known which are stable only in such solutions, Raman spectrometry is the method of choice for their study.

2.4.2 *Highly Hydrogen Bonded Systems*

The widest use of Raman spectrometry in this area is for the study of aqueous solutions. Although water is a corrosive liquid as far as most infrared cell materials are concerned, its most troublesome feature is its relative opacity to infrared radiation. Water is almost the ideal solvent for Raman spectrometry, however, as it has only a few minor features in a normal Raman spectrum. As a result, many ionic species which exist only in aqueous solution can be readily studied. Figure 7 exemplifies the use of water as a solvent by reproducing the Raman spectrum of dimethylsulfone in aqueous solution. This material is essentially insoluble in all normal organic solvents which might be used for infrared spectral studies.

Perhaps the most promising area of aqueous solutions for study is that of biologically interesting molecules. Many of these compounds are extremely complex and very sensitive to their environment. Since they nearly always exhibit their interesting biological consequences in an aqueous media and often degrade or irreversibly change upon dehydration, they must be studied in the presence of water if reliable conclusions are to be drawn. Raman spectra have been shown to be sensitive to the conformation of biopolymers and have been used to study local areas of a material in a matrix of some other material, by utilization of the laser's property of very small beam divergence.

A further use is found in that many biologically significant molecules have a large degree of hydrogen bonding which tends to obscure their infrared spectra. In general, their Raman spectra are of considerably higher quality and other structural features, which can be obscured in their infrared spectra, become quite apparent. This effect has previously been mentioned with respect to sugars and other carbohydrates.

2.4.3 *Polymers*

There has recently been a great deal of interest in the application of Raman spectrometry to the field of polymer characterization. From the detailed structural point of view, the Raman spectra of polymers are generally more sensitive to polymer conformation than are their infrared spectra. This is because most polymers have a backbone of carbon–carbon bonds while their side chains contain polar groups. Since Raman spectra are more sensitive to the various vibrations of the carbon chain they reflect its nature to a much greater extent. In general, the Raman spectrum of a polymer

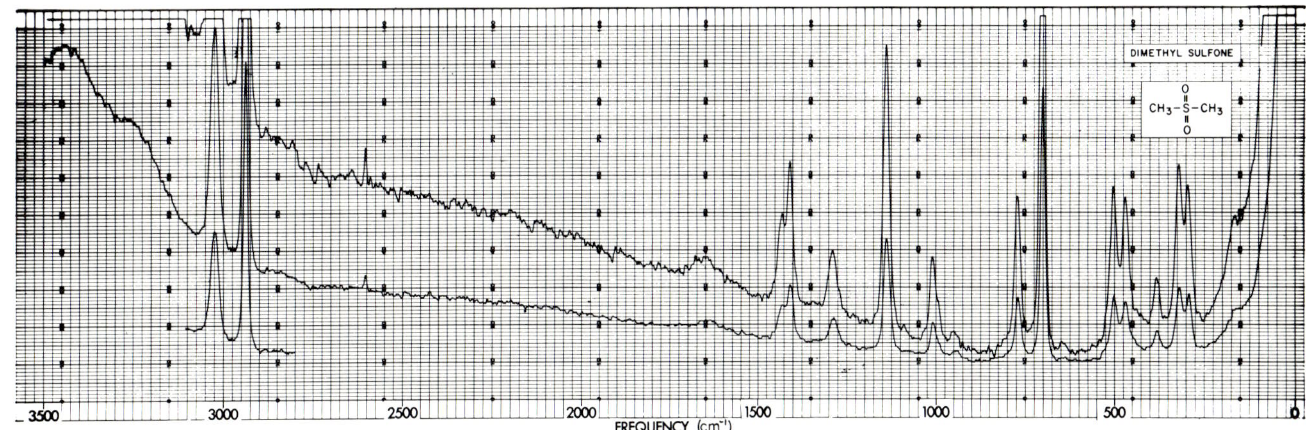

Fig. 7. The Raman spectrum of a 30% aqueous solution of dimethylsulfone. The strong broad band just below 3500 cm^{-1} and the weak, broad band near 1650 cm^{-1} are due to water. The very strong band at 700 cm^{-1} is due to the symmetric CSC stretching mode and the somewhat weaker band at 1138 cm^{-1} is due to the symmetric SO_2 stretching mode. [From Feairheller and Katon, 1964.]

is quite sensitive to its conformation (extended versus helical) and tacticity.

In a more practical sense, identification of a given polymer by infrared methods is usually tedious. Most polymers cannot be ground into small particles and therefore the only way a good spectrum can be obtained is by dissolving and then casting a film. The bulk polymer is readily studied by Raman spectrometry, however, with practically no sample preparation time required. In addition, the Raman spectra are often more useful than the infrared for indicating certain chemical properties of the polymer such as number of S—S linkages in vulcanized rubber, number of C=C linkages in polyolefins, etc.

2.4.4 *Inorganic Materials*

Raman spectrometry has long been a preferred method for studying many inorganic materials. Formerly, it found its major advantage in the ease with which low frequency spectra could be measured since most inorganic vibrations are at low frequency. It has several other advantages over infrared and other methods, however.

One of the great advantages of using the laser as an excitation source is its very narrow beam width. This makes it possible in many cases to obtain the Raman spectra of small regions of a mineral sample and thus study one mineral occluded in another in a nondestructive manner.

Many minerals are refractory and extremely difficult to prepare for infrared studies. Both powders and large crystals are readily investigated by Raman techniques, however. In the area of single crystals Raman has very distinct advantages since there is no size limitation. In most cases, it is difficult to prepare a single crystal sufficiently thin to obtain a good infrared spectrum however.

The glasses have been studied for many years by Raman spectrometry since they are usually opaque to infrared radiation. With the advent of laser Raman spectrometry much higher quality spectra are now obtainable. In fact, Raman spectrometry is one of the few techniques applicable to identification of these relatively intractable materials.

There has been a good deal of work on the Raman spectra of inorganic compounds which takes advantage of another useful feature of Raman spectrometry. This is the area of study of the structures of fused salts and high-temperature melts which often contain unusual ionic species. Raman spectra can be readily recorded on liquid samples at temperatures of the order of 1000°C. Such studies in the infrared are impossible or very difficult because the sample radiation is nearly as great as that of the source in the infrared region. This leads to very poor spectra.

3 Data Interpretation

The interpretation of Raman spectral data is very similar to that of infrared data, as one would expect since the two both measure vibrational characteristics of the material under investigation. In the area of qualitative analysis, the "fingerprint" technique is just the same in the two methods—namely, the matching of an unknown spectrum to that of a known. In the area of structure determination, account is taken of the symmetry of the molecule. This sort of treatment can be found in any of the texts on theoretical treatment of vibrational spectra. Both infrared and Raman spectra are treated simultaneously by these symmetry techniques.

The group frequency concept is similar to that used in the infrared, but Raman group frequencies usually have different intensity features than do their infrared counterparts. The best compilation of Raman data available at present is probably that of Tobin (1971), although Hibben (1939) also has a compilation which is useful (although much older). As mentioned previously, any good infrared spectral correlation chart is useful for Raman correlations as well. Raman intensities will be quite different from those for infrared absorption, but intensities are seldom listed in correlation charts in any case. The reason the correlation charts are useful for both is that both techniques measure exactly the same vibrational frequency of the molecules in question as long as the molecules do not possess very much symmetry.

Annotated References

Biennial reviews in *Anal. Chem.* Reviews which summarize important developments and applications of Raman spectrometry as they apply to analytical chemistry.

Colthup, N. B., Daly, L. H., and Wiberley, S. E. (1968). "Introduction to Infrared and Raman Spectroscopy." Academic Press, New York. An elementary treatment for analytical chemists which compares infrared and Raman spectroscopy.

Feairheller, W. R., Jr., and Katon, J. E. (1964). *Spectrochim. Acta* **20,** 1099.

Gilson, T. R., and Hendra, P. J. (1970). "Laser Raman Spectroscopy." Wiley (Interscience), New York. A somewhat more theoretical treatment than is given by Tobin. Of particular note is the coverage of single crystals, gases, and polymers.

Herzberg, G. (1950, 1945). "Spectra of Diatomic Molecules" and "Infrared and Raman Spectra of Polyatomic Molecules." Van Nostrand Reinhold, Princeton, New Jersey. General theoretical treatment of vibrational spectroscopy including aspects of molecular symmetry.

Hibben, J. H. (1939). "The Raman Effect and Its Chemical Applications." Reinhold, New York. A full account of the early work in Raman spectroscopy. Although somewhat dated, the theory is covered in some detail and results up until that time are fully discussed.

Katon, J. E., and Carll, T. P. (1971). *J. Mol. Structure* **7,** 391.

Katon, J. E., and McDevitt, N. T. (1965). *Spectrochim. Acta* **21,** 1717.
Katon, J. E., Feairheller, W. R., Jr., and Lippincott, E. R. (1964). *J. Mol. Spectrosc.* **13,** 72.
Kiefer, W., and Bernstein, H. J. (1971). *Appl. Spectrosc.* **25,** 500, 609.
Koenig, J. L. (1971). *Appl. Spectrosc. Rev.* **4,** 233. A review on the Raman spectra of polymers.
Rippon, W. B., Koenig, J. L., and Walton, A. G. (1971). *J. Agr. Food Chem.* **19,** 692. A review of the application of Raman spectrometry to biopolymers and proteins.
Rosenbaum, E. J. (1965). Raman spectroscopy. *In* "Treatise on Analytical Chemistry" (I. M. Kolthoff and P. J. Elving, eds.), Part 1, Vol. 6. Wiley (Interscience), New York. A rather short treatment of prelaser Raman spectroscopy notable primarily for the fact that it has a section on quantitative analytical application.
Sloane, H. J. (1971a). *Appl. Spectrosc.* **25,** 430. An excellent up-to-date comparison of all aspects of Raman and infrared spectroscopy.
Sloane, H. J. (1971b). The use of group frequencies for structural analysis in the Raman compared to the infrared. *In* "Polymer Characterization" (C. D. Craver, ed.). Plenum Press, New York. A good review of group frequencies and their Raman characteristics for both simple molecules and polymers.
Szymanski, H. A. (ed.) (1967, 1970). "Raman Spectroscopy, Theory and Practice," Vols. I and II. Plenum Press, New York. A two volume collection of review articles on specific applications and techniques written by experts. Many applications to various fields of chemistry are covered as well as specialized techniques for unusual studies.
Tobin, M. C. (1971). Laser Raman spectroscopy. *In* "Chemical Analysis" (P. J. Elving and I. M. Kolthoff, eds.), Vol. 35. Wiley (Interscience), New York. An excellent, up-to-date text written primarily for the chemist. It contains a brief section on the theory, a good section on experimental details, and a chapter on group frequencies in which the infrared and Raman spectral characteristics of many vibrations are compared. It also has a brief chapter on applications to various fields of chemistry and a bibliography arranged according to subject matter.

CHAPTER 15

Refractometry

J. H. Richardson

The Aerospace Corporation
El Segundo, California

Introduction

Refractometry or the measurement of refractive indices is a rapid and precise confirmatory test in the qualitative analysis of gases, liquids, and translucent solids. The measurement of refractive index also finds a major application in the quantitative analysis of single-phase mixtures.

For the purpose of this chapter, materials may be divided into two groups: (1) isotropic materials, that is, materials with only one refractive index for a single wavelength of light; and (2) anisotropic materials, that is, materials with more than one refractive index for a single wavelength. These two classes of materials are categorized in Table 1.

This chapter deals only with isometric materials. Measurements of the refractive indices of anisotropic materials are discussed in Chapter 27.

TABLE 1

OPTICAL CLASSIFICATION OF MATERIALS

Class			Number of refractive indices[a]
Isotropic	(a)	Gases and their mixtures	1
	(b)	Liquids, mixtures and solutions	1
	(c)	Isotropic crystalline solids and their solid solutions, glasses, and noncrystalline polymers	1
Anisotropic	(a)	Uniaxial crystalline solids (hexagonal and tetragonal) and their solid solutions	2
	(b)	Biaxial crystalline solids (orthorhombic, monoclinic, and triclinic) and their solid solutions	3

[a] These numbers represent the refractive indices for a single wavelength of light.

1 Theory

1.1 Refractive Index

Refraction is the bending of a beam of light upon passing from one medium to another. The refractive index is a measure of this effect and is defined as the ratio of the velocity of light in vacuum to the velocity of light in the material, i.e.,

$$\eta_m = V_v/V_m \qquad (1)$$

where η_m is the refractive index of m, V_v is the velocity of light in vacuum, and V_m is the velocity of light in m.

The familiar Huygens wavefront construction may be used to account for the angular relationships; these relationships are illustrated in Fig. 1. From these constructions may be deduced the general expressions relating the refractive indices, velocities of light, and angular relationships between the beams of light incident on and refracted from the interface between two phases of different refractive index.

$$\eta_2/\eta_1 = V_1/V_2 = \sin \phi_1/\sin \phi_2 \qquad (2)$$

When a beam of light passes from a higher to a lower index medium, it is bent away from the normal to the interface; see, for example, a beam passing

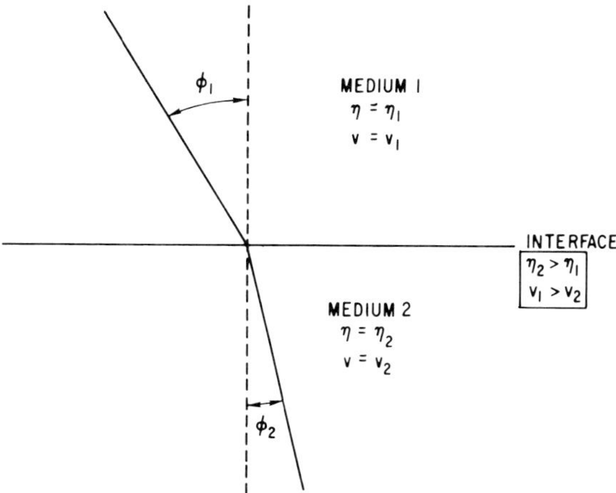

FIG. 1. Angular relationship between beams of light incident on and refracted from an interface between two phases with different refractive indices.

from medium 2 to medium 1 in Fig. 1. If the angle of incidence, in this case ϕ_2, is increased sufficiently, the angle of refraction η_1 will become 90° and the refracted ray will travel parallel to the interface. That value of the angle of incidence for which the angle of refraction is 90° is known as the critical angle. If ϕ_2 is further increased, the beam of light no longer enters the lower index medium; this is known as total reflection.

The actual value of the refractive index depends on the temperature or the pressure on the refracting medium as well as on the wavelength of the light. Properties of the material also affect the refractive index; these include the chemical composition, physiochemical constitution, state, purity, and homogeneity.

1.2 DISPERSION

Because the refractive index of a material depends on the wavelength of the light, most refractive indices are given for a single wavelength. The wavelength usually employed is the "D" line (589.262 nm), which is the weighted average of the D_1 (589.593 nm) and the D_2 (588.996 nm) lines of the sodium spectrum.

The refractive index of all common materials increases as the wavelength of light decreases; this effect is known as dispersion. Because of this dispersion, a glass prism bends blue light more than red, thus spreading white light into the characteristic spectrum of its component colors.

A number of expressions have been used to give a value for the dispersion of a material. The dispersive power w of a material is given by

$$w = (\eta_f - \eta_c)/(\eta_d - 1) \tag{3}$$

where η_f is the refractive index with 486.1-nm light, η_c is the refractive index with 656.3-nm light, and η_d is the refractive index with 589.3-nm light.

In the optical industry the Abbe number of ν value is used where

$$\nu = 1/w \tag{4}$$

2 Instrumentation and Methods

2.1 Gases

Because the low refractive indices of gases require instruments of high sensitivity, interferometric instruments are used for gas refractometry.

The heart of an interference-type refractometer is two identical gas cells. These cells are simultaneously illuminated by an optical arrangement, called a beam splitter, that separates a partially coherent beam of light into two beams. When these beams have traversed the two cells, they are recombined into a single beam by a second optical arrangement, a beam splitter in reverse. An interference refractometer is shown schematically in Fig. 2.

When the recombined beam of light is examined, a system of interference fringes is observed. If the interferometer is illuminated with white light, a colorless fringe is formed that is bordered on either side with alternating dark and light fringes. These outer fringes are tinted red on the edge opposite the central fringe. The difference in path length for the two beams passing through the two cells for each bright fringe observed is $\lambda/2$, where λ is 550.0 nm, the average wavelength of the white light.

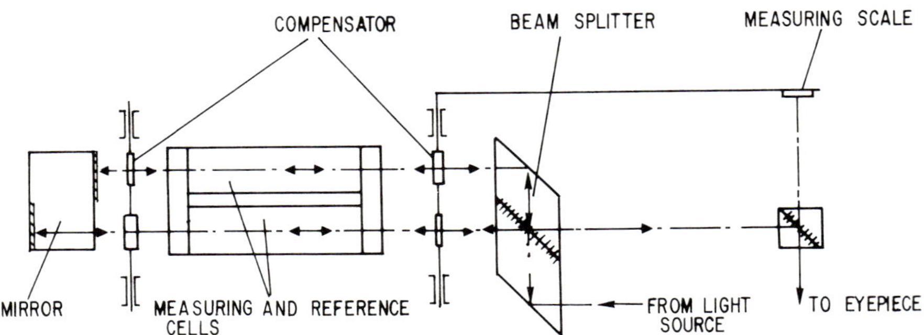

Fig. 2. Schematic drawing of an interference refractometer. [From Kinder and Torge, 1969. Reprinted courtesy of Carl Zeiss.]

The absolute refractive index of a gas may be determined in the following manner. The sample and reference cells are evacuated, and the position of the central fringe is noted. The gas is slowly allowed to fill the sample cell while the fringe pattern is observed. The total fringe movement N in units of fringe spacing is noted, usually at conditions of standard temperature and pressure, 0°C and $1.013 \times 10^5 N/m^2$ (760 Torr).

The refractive index is given by

$$\eta = (N\lambda/L) + \eta_0 \tag{5}$$

where N is the fringe displacement in units of fringe spacing, λ is the wavelength of the illumination (550.0 nm), and η and η_0 are the refractive indices of the sample and the standard, respectively.

Monochromatic light is not normally used for these measurements, since the characters of the bright central fringe and the tinted outer fringes as viewed in white light permit easy identification of the fringes during their movement.

It should be emphasized at this point that, although the refractive indices of the gases are well known and can be measured with accuracy, this method is not ordinarily used as a means of gas identification. The interferometric refractometer does find application, however, in the quantitative analysis of gas mixtures. The resultant refractive index r of a mixture composed of k components each of concentration V (volume percent) may be calculated as follows:

$$\eta = \sum_{i=0}^{k} (V_i/100)\eta_i \tag{6}$$

where V_i is the volume percent and η_i is the refractive index of the ith component.

Thus the refractive index of a binary mixture of a gas with the refractive index η_1 that contains V_2 of a second gas with a refractive index η_2 is given by

$$\eta = (V_2/100)\eta_2 + [(100 - V_2)/100]\eta_1 \tag{7}$$

One method of analysis consists of filling the reference cell with the pure major gas and the sample cell with the mixture. Substitution in Eq. (5) with rearrangement gives

$$N = (L/\lambda)(\eta - \eta_0) \tag{8}$$

Combining with Eq. (7) gives

$$N = (L/\lambda)(V_2/100)(\eta_2 - \eta_1) \tag{9}$$

and

$$V_2 = N\lambda 100/(\eta_2 - \eta_1)L \tag{10}$$

Gas refractometry has been used in quantitative analysis of the methane or other hydrocarbons in mine gases (Kinder, 1956). In this application, the sample cell is filled with the atmosphere of the mine, which is measured against a reference gas. With the Zeiss Fire Damp interferometer, concentrations of 5% (the lower limit of explosion) and less may be observed for methane.

A second application for gas refractometry has been the calibration of anesthetic evaporators (Kinder and Torge, 1969). In this application, the concentration of fluothane gas can be easily measured in the range from 0 to 5%.

The range of an interference refractometer is inversely related to the length of the interferometer cells. Characteristics of typical interferometers suitable for gas analysis are shown in Table 2.

TABLE 2

Refractometer Characteristics

Phase to be measured	Type of refractometer	Sample size	Range (η)	Precision (η)
Gas	Interferometer	25–100 cm[a]	±0.00005[b]	$\pm 1 \times 10^{-8}$
Liquid	Abbe	0.01–0.10 ml	1.17–1.87[c]	$\pm 1 \times 10^{-4}$
	Precision Abbe	0.05–0.10 ml	1.20–1.70[c]	$\pm 3 \times 10^{-5}$
	Dipping	0.1–30 ml	1.33–1.64[d]	$\pm 1 \times 10^{-5}$
	Pulfrich	0.5–5 ml	0.05[b]	$\pm 1 \times 10^{-5}$
	Pulfrich	0.2–5 ml	1.33–1.86	$\pm 1 \times 10^{-4}$
	Prism spectrometer	0.04–25 ml	Wide	$\pm 1 \times 10^{-6}$
	Jelly–Fisher	0.00001–0.05 ml	1.30–1.90	$\pm 1 \times 10^{-3}$
	Nichols[e]	0.1 ml	1.30–2.0	$\pm 1 \times 10^{-3}$[f]
	Interferometer	4 cm[a]	0.0006[b]	$\pm 1 \times 10^{-7}$
Solid	Abbe	35–40 mm[g]	1.17–1.87	$\pm 2 \times 10^{-4}$
	Abbe	6 mm^2 [h]	1.17–1.87	$\pm 2 \times 10^{-4}$
	Pulfrich	40–120 mm^2	1.33–1.86	$\pm 1 \times 10^{-4}$
	Pulfrich	50 mg[i]	1.33–1.86	$\pm 2 \times 10^{-4}$
	Immersion[e]	5-μm diam	1.33–2.8	$\pm 1 \times 10^{-3}$
	Prism spectrometer	Large polished prism	Wide	$\pm 1 \times 10^{-6}$

[a] Path length.
[b] Differential measurement.
[c] Complete range requires interchange of prisms.
[d] Complete range requires 10 prisms.
[e] Used in conjunction with an optical microscope.
[f] Can be increased to 5×10^{-4} by use of monochromatic light.
[g] Area of sample surface.
[h] Areas as small as 6 mm^2 may be measured by method of Maechler and Kortz (1966).
[i] 20–40 mesh powder by the method of LeBlanc (1892).

2.2 Liquids

The major application of refractometry is in the analysis of liquids. Many instruments are available for the measurement of refractive index of liquids; however, these have significantly different sensitivities and ranges. These instruments generally employ (1) the critical angle phenomenon, (2) image displacement, or (3) interference phenomena such as described above.

Instruments, such as the Abbe, Pulfrich, and dipping refractometers, make use of the critical angle phenomenon. The Abbe refractometer will be discussed in some detail as an illustration of this type. Information on the other types may be found in Bauer *et al.* (1960) and Tilton and Taylor (1950). Characteristics of these are listed in Table 2.

The Abbe refractometers are manufactured in one of two ranges of sensitivity, one for routine work and the other for more precise measurements. The operation of a rather uniquely arranged Abbe refractometer is shown in Fig. 3.

In operation, prism B is raised and the liquid sample, ranging in volume from 0.1 to 1.0 ml, is placed on the surface of prism A. Prism B is closed,

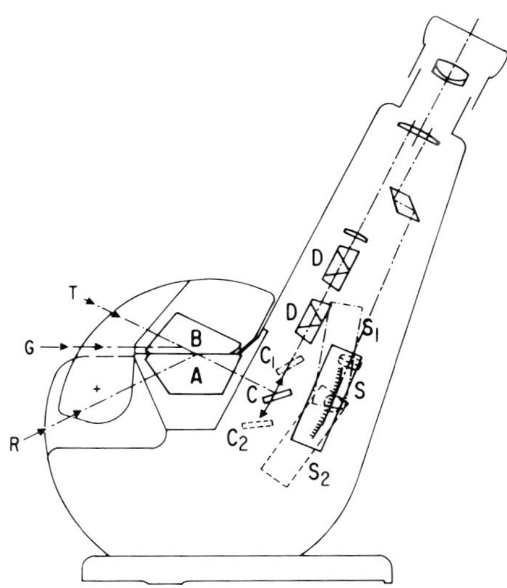

Fig. 3. Schematic drawing of an Abbe refractometer of recent design. A is the measuring prism and B the auxiliary prism. C and S, C_1 and S_1, and C_2 and S_2 show the positions of the collecting mirror and scale for three different refractive index measurements. D is an Amici prism. G, R, and T are the paths of the illumination for grazing incidence, reflection, and transmission, respectively. [From Zeiss Abbe-Refractometer Instructions G 50-110/11c. Reprinted courtesy of Carl Zeiss.]

and the instrument is illuminated with white light from the direction T. For transparent liquids, the illumination enters from the left side of prism B, passes through the liquid, and then passes through prism A. Opaque and colored liquids may be examined in a reflection mode by illumination of the sample and prism A from below; the path of the illumination in this case is from direction R.

Adjustment of a large knob on the side of the instrument moves mirror C and the scale S simultaneously, thereby locating the area of the critical angle in the visual image. When the critical angle is in view, there appears in the upper portion of the image, as shown in Fig. 4, a bright field above and a dark field below; these fields are separated by a boundary, which is usually surrounded by colored fringes. The colored fringes are removed by rotation of one of the Amici prisms D by use of a calibrated knob on the side of the instrument. The boundary is then adjusted to the intersection of the cross hairs, and the index of refraction is read from the scale and cross hair in the lower portion of the image.

The dispersion of the liquid may be determined from the value shown by the calibrated knob that rotates the Amici prism by use of the expression

$$\eta_f - \eta_c = A + kBS \tag{11}$$

where A and B are functions of dispersion, S is a function of the Amici prism setting, and k is an instrument constant. A nomograph combining each of these factors is supplied with the Abbe instrument described herein.

The Amici prisms achromatize the instrument and adjust the refractive index reading for the sodium D line (589.3 nm). While this is desirable for routine work, it should be emphasized that this feature is not suitable for a refractometer to be used for other wavelengths of light, for example, in a double-variation measurement of refractive index, see Emmons (1943). Instruments without the Amici prisms are available for this purpose.

To realize the full accuracy of the refractometer, it is necessary to

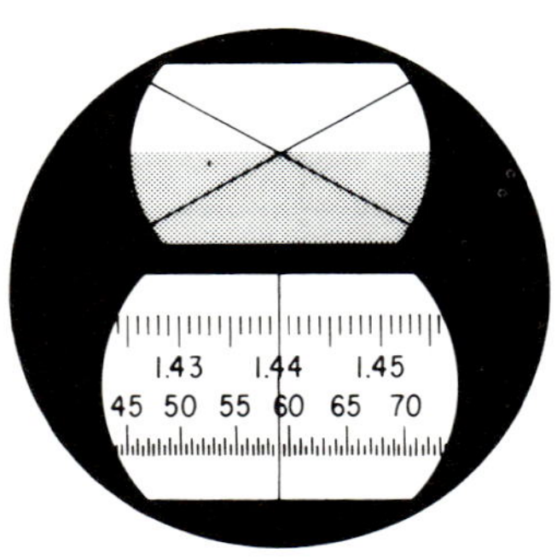

FIG. 4. The field of view in the Abbe refractometer shown in Fig. 3. In this instrument, the image of the boundary and the refractive index scale are seen simultaneously. [From Zeiss Abbe-Refractometer Instructions G 50-110/11c. Reprinted courtesy of Carl Zeiss.]

control the temperature to ±0.2°C. Therefore, the Abbe instrument described is provided with connections for recirculating water. The actual temperature selected for the measurement of the sample should correspond to the temperature employed for the measurement of the knowns. Unfortunately, the temperatures for refractive indices listed in the various literature references, e.g., the *Handbook of Chemistry and Physics*, range from 17 to 23°C, but unless otherwise specified, the refractive indices given are for 20°C.

Sylvester and Houlihan (1962) have described a simple assembly to provide circulating water of a desired temperature for use with a refractometer. Such a system will permit exact matching of the temperature of the material with that used in the literature.

A second optical effect which may be used in the determination of the refractive index of a material is image displacement. This effect is used in both the spectrometer method and the Jelly–Fisher refractometer. The characteristics of these instruments are given in Table 2.

A hollow prism and a source of monochromatic light are required for the determination of the refractive index of a liquid in a spectrometer. The prism may be a hollow prism filled with the liquid or it may consist of only two sides of the prism formed by cementing two microscope cover slips together, in which arrangement a single drop of the sample usually remains at the junction of the two cover slips long enough for measurement (Larsen and Berman, 1934).

The spectrometer is used to find the angle of minimum deviation for a beam of monochromatic light through the sample prism; this is shown schematically in Fig. 5. The refractive index of the liquid is then

$$\eta = \frac{\sin(D + A)/2}{\sin A/2} \tag{12}$$

where D is the angle of minimum deviation and A is the prism angle.

Larsen and Berman suggest the use of a 60° prism for moderate-index liquids and a 30° prism for high-index liquids. Characteristics of the spectrometer are given in Table 2.

The final technique for the measurement of the refractive index of liquids is the interferometric method. This method is described in the section on gases. Since this is a differential method, a standard is required; this should be a liquid of accurately known refractive index. The characteristics of the interferometric technique for liquids are listed in Table 2.

2.2.1 *Qualitative Analysis of Liquids*

The measurement of refractive index and dispersion is usually inadequate for unambiguous identification of a liquid even if it is pure. There-

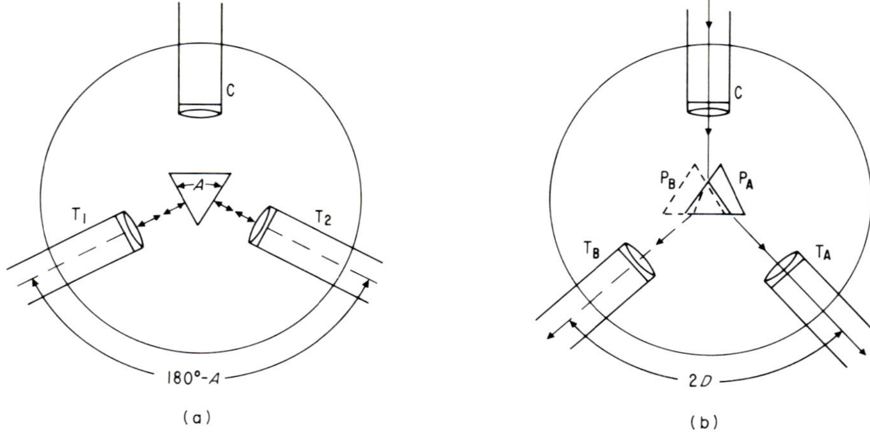

FIG. 5. Schematics showing the use of the prism spectrometer to measure refractive indices. (a) The prism angle A is measured with the use of the telescope as an autocollimator at positions T_1 and T_2. The collimator C is not used for this measurement. The angle between T_1 and T_2 is 180°—A. (b) Measurement of angle of minimum deviation. Light from C passes through the prism and is observed at T_A when the prism is in position P_A and at T_B when the prism is in position P_B. The angle between T_A and T_B is $2D$. [Reprinted from Kingslake (1967), courtesy of Academic Press, Inc.]

fore, these measurements are generally used only in a confirmatory role in conjunction with other analytical methods.

Even when these measurements are used in this confirmatory role, two rather different applications may be made of the data. In the first, the refractive index and the dispersion values may be compared with literature values of suspected compounds. Agreement of these values provides additional confirmation for the identification.

The second application is suitable if the refractive indices of the possible compounds are not in the literature. A knowledge of the chemistry of the suspected compounds may be employed to calculate their molecular refraction.* Bauer *et al.* (1960) discuss this procedure in detail and list values for the various types of contributions. The refractive indices of all

* The molecular refraction R of a compound is given by the Lorentz–Lorenz equation

$$R = (\eta^2 - 1)/(\eta^2 + 2)\,(M/\rho)$$

where M is the molecular weight and ρ is the density of the material. R is constant only for very long wavelengths of light; however, it may be compared for various molecules at a given wavelength. Since each atom, bond, or group contributes to the overall refractive index, R may be computed additively from the atomic or bond refractive values.

suspected compounds with the empirical formula of the unknown may then be calculated from their molar refraction and densities using the rearranged Lorentz–Lorenz equation

$$\eta = [(M + 2R\rho)/(M - R\rho)]^{1/2} \tag{13}$$

The refractive index of the material may then be compared with the indices calculated for the suspected compounds.

2.2.2 *Quantitative Analysis of Liquids*

The refractive index may be used as a measure of the concentration of either solids or liquids in liquids. Indeed, some of the commercial refractometers may have, in addition to the scale for refractive index, a scale for the determination of the concentration of solid solutes, e.g., sugar, or liquid solutes, e.g., butterfat. Caution should be exercised in the use of these scales, since the presence of unsuspected solutes may cause large errors in the results.

The most reliable method for the determination of the amount of solute is interpolation from an empirically determined calibration curve. Since the components may undergo association in solution, a linear relationship does not usually exist in the curve for η versus concentration, C. A few curves have been published for selected systems; major references are given in the Selected Reading list.

Expressions for η as a function of C have been derived by many, e.g., Joshi (1963) and Glover and Goulden (1963). These are applicable only in certain systems and concentration ranges.

2.3 Solids

The measurement of refractive index has been important for a number of years in the identification of nonopaque materials. The major contribution to this technique has been made by the mineralogists, who have prepared determinative tables for as many as 1200 nonopaque minerals.

Solids may be conveniently grouped into two broad categories: (1) the isotropic materials, which have only one refractive index, and include cubic crystalline solids, amorphous materials, and glasses; and (2) the anisotropic materials, which have two or three refractive indices, and include all noncubic crystalline solids.

Since the refractive indices of isotropic solids may be measured on many of the same instruments used for liquids and gases, they are discussed in this chapter. Methods for the anisotropic materials, which are significantly more involved, are discussed in Chapter 27, which deals with optical microscopy.

2.3.1 *Qualitative Analysis of Solids*

If one is afforded the luxury of a large transparent sample that may be ground and polished into a triangular prism,* the refractive index may be determined easily and with high accuracy, as shown in Table 2, using the prism spectrometer.

The refractive index of solid materials may also be obtained in grazing incident light with either the Abbe or the Pulfrich instrument. This also requires the preparation of two perpendicular, planar, polished surfaces that intersect at a sharp edge. The larger of these surfaces should be at least 35 to 40 mm^2. The arrangement of the surfaces is shown schematically in Fig. 6. The sample is affixed by its major polished surface to the measuring prism of the refractometer with a small droplet of a liquid that does not react with the sample and has a greater refraction than that of the sample; monobromonaphthalene is usually used. The specimen is rotated on this surface until its minor surface is normal to the illumination. The grazing incidence illumination path G is illustrated for the Abbe refractometer in Fig. 3. The illumination prism remains in an elevated position for the examination of solids and serves only as a shade for extraneous light. The refractive index of the solid is determined in the same manner as that used for liquids.

A solid may also be measured in reflected light in the Abbe refractometer as shown in Fig. 3. Illumination path R is used in this case. Only one polished surface is required for this measurement; this surface must be at least 35 to 40 mm^2 unless special steps are taken to increase the contrast of the image seen in the refractometer.

Maechler and Kortz (1966) have described a method for examining polished surfaces as small as 6 mm^2. In this method, light that would normally be reflected from the portions of the measuring prism not in contact with the sample is minimized by use of a paste to cover the meas-

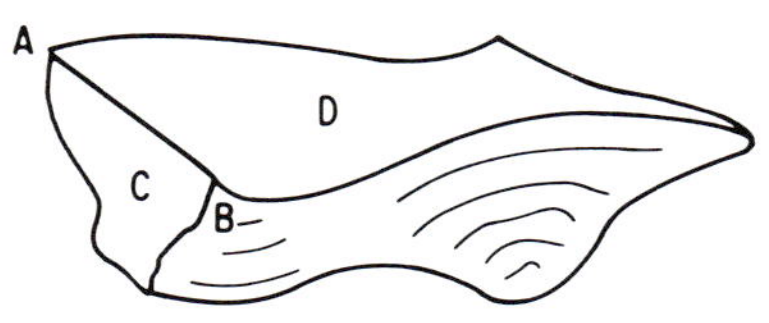

FIG. 6. Arrangement of ground and polished surfaces C and D on a solid material for measurement of refractive index in grazing incident light with an Abbe refractometer. The two surfaces are at right angles and meet in a sharp line AB. For measurement of refractive index in reflected light on the Abbe refractometer, only one surface, D, is required.

* At least two planar, polished surfaces are required. A 60° wedge angle between the two surfaces is desirable for materials of moderate refractive index; a 30° angle may be better for materials of high index.

uring prism around the specimen. This paste consists of fine carbon black dispersed in a liquid having the same refractive index as the measuring prism.

The dispersion of the solid may be determined on the Abbe refractometer in the same manner as that used for liquids.

The third method for the determination of the refractive index of an isotropic solid employs an optical microscope* and a calibrated set of refractive index oils.† This method may use discrete particles from approximately 5 to 500 μm. Samples of larger size should be crushed and sieved; the 100 mesh (149 μm) fraction is suitable for this purpose.

In practice, the particles are immersed in one of the refractive index oils on a glass microscope slide and covered with a cover slip. The mounted particles are then examined under the optical microscope, and the refractive index of the sample is compared with that of the oil.

Three methods of comparing the refractive indices of the material and the surrounding oil are discussed by Saylor (1935). The Becke line method is discussed here, since this method requires no ancillary equipment for the microscope and is capable of matching the indices with a precision of $\pm 1 \times 10^{-3}$.

The objective lens of the microscope should be chosen such that the sample fragments are comfortably visible; usually 4× to 45× objectives are most suitable. The aperture diaphragm should be adjusted to give a narrow cone of illumination. The slide should be placed on the stage and critically focused. As the objective lens is moved away from the position of critical focus, a bright line, the Becke line, is seen to move away from the interface to the medium of higher index. On the basis of this observation, another slide is prepared using a second oil, the refractive index of which approaches the index of the sample. This process is repeated until the refractive index of the material is matched. For monochromatic light, this match obtains when the Becke line disappears and the outline of the fragment vanishes.

Once the refractive index of the fragment has been determined it may be compared with the literature values.

The dispersion of the material may be determined by measurement of the refractive index at different wavelengths. This may be accomplished by use of a mercury lamp with appropriate filters or by use of a wedge interference filter such as that described by Jones (1965). It should be emphasized that there is generally less information available on the dispersion than there is for the η_d of solids.

* A standard biological microscope is adequate for this purpose.

† Available from R. P. Cargille Laboratories, Inc., Cedar Grove, New Jersey 07009.

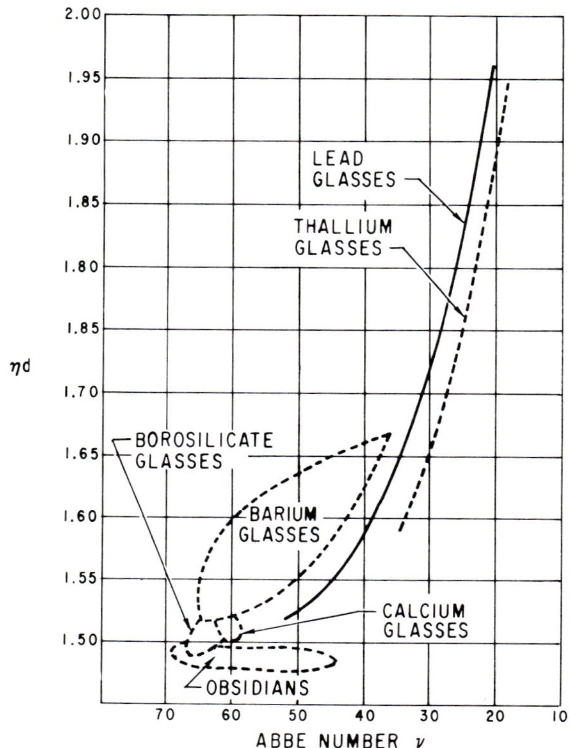

FIG. 7. Utility of refractive index measurements in the qualitative analysis of natural and manufactured glasses. Measurement of the refractive index η_d and the Abbe number ν permits the identification of the glass type. [Reprinted from Winchell and Winchell (1964) courtesy of Academic Press, Inc.]

Wright (1920) has developed a useful correlation between the refractive index and the dispersion to aid in the identification of natural and manufactured glasses. A graph of such data is shown in Fig. 7. See the Selected Reading list for further references.

2.3.2 *Quantitative Analysis of Solids*

Analyses of binary solid solutions by use of refractometry are possible; however, these depend on empirically determined curves of η versus C. Examples of such curves are limited, but data on selected systems are to be found in references listed in the Selected Reading list.

References

Bauer, N., Fajans, K., and Kewin, S. Z. (1960). *In* "Physical Methods of Organic Chemistry" (A. Weissberger, ed.), Part II, pp. 1139–1281. Wiley (Interscience), New York.

Emmons, R. C. (1943). "The Universal Stage," Memoir 8, pp. 55–101. Geol. Soc. Amer., New York.

Glover, F. A., and Goulden, J. D. S. (1963). *Nature (London)* **200** (4912), 1165–1166.

Jones, F. T. (1965). *Microsc. Crystal Front* **14** (11), 440–447.

Joshi, D. P. (1963). *J. Indian Chem. Soc.* **40** (7), Parts I and II, 550–554.

Kinder, W. (1956). *Zeiss Inform.* No. 19, 19–24.

Kinder, W., and Torge, R. (1969). *Zeiss Inform.* No. 69, 80–85.

Kingslake, R. (1967). "Applied Optics and Optical Engineering," Vol. IV, Part I, p. 367. Academic Press, New York.

Larsen, E. S., and Berman, H. (1934). The Microscopic Determination of The Nonopaque Minerals, pp. 18–20. U.S.G.S. Bull. 848, U.S. Dept. of The Interior, Washington, D.C.

LeBlanc, M. (1892). *Z. Phys. Chem.* **10,** 433.

Maechler, M., and Kortz, A. (1966). *Zeiss Inform.* No. 58, 154–155.

Saylor, C. F. (1935). *J. Res. NBS* **15,** 272–294.

Sylvester, H. M., and Houlihan, W. J. (1962). *J. Chem. Ed.* **39** (5), 538.

Tilton, L. W., and Taylor, J. K. (1950). *In* "Physical Methods in Chemical Analysis" (W. G. Berl, ed.), Vol. I. Academic Press, New York.

Winchell, A. N., and Winchell, H. (1964). "The Microscopical Characters of Artificial Inorganic Substances," p. 324. Academic Press, New York.

Wright, F. E. (1920). *J. Amer. Ceram. Soc.* **3,** 783.

Selected Reading

Sources of Refractive Index Information

Batsanov, S. S. (1961). "Refractometry and Chemical Structures." Plenum Press, New York.

Kirk, P. L. (1951). "Density and Refractive Index, Their Application in Criminal Identification." Thomas, Springfield, Illinois.

Lange, N. A. (1967). "Handbook of Chemistry." McGraw-Hill, New York.

Timmermans, J. (1950, 1965). "Physico-Chemical Constants of the Pure Organic Compounds," Vols. I and II. Elsevier Publ., Amsterdam.

Washburn, E. W. (1930). "International Critical Tables," Vol. VII. McGraw-Hill, New York.

Weast, C. (ed.) (1971). "Handbook of Chemistry and Physics." Chem. Rubber, Cleveland, Ohio.

Winchell, A. N. (1954). "The Optical Properties of Organic Compounds." Academic Press, New York.

Winchell, A. N., and Winchell, H. (1951, 1956). "Elements of Optical Mineralogy," Parts II and III. Wiley, New York.

Winchell, A. N., and Winchell, H. (1964). "The Microscopical Characters of Artificial Inorganic Solid Substances." Academic Press, New York.

CHAPTER 16

Scanning Electron Microscopy

John C. Russ

EDAX Laboratories
Raleigh, North Carolina

Introduction

Because the scanning electron microscope (SEM) is a relatively new device, only available commercially within the past several years, its general use for many types of problems, including analytical ones, is not yet fully explored or documented. This chapter will try, therefore, to give the reader a broad idea of the SEM, its more general or "qualitative" uses, and the ability of the instrument to provide various types of quantitative analysis. Since little documentation exists in this latter category, some of the numbers presented for sensitivity, etc., will of necessity be conservative estimates, and subsequent developments in the use of SEMs will most likely result in substantial improvement of these values.

The SEM is quite different in principle from the conventional transmission electron microscope (TEM), and much more recently arrived on the commercial scene, but its history goes back just as far. Knoll, in Germany, suggested in 1935 the possibility of a device very similar to the present day SEM, and the scanning principle was employed in an operating instrument three years later by von Ardenne. A more practical SEM that demonstrated 500-Å resolution was built at RCA by Zworykin and Hillier in 1940, but the war halted further work. After the war, research went on principally in Europe, and especially at Cambridge by Oatley and his co-workers (1965). The efforts of the latter group led directly to the first successful commercial instrument, the Stereoscan, in 1965. Since then, other instruments have come onto the market, made in Japan, France, and the United States. By the end of 1971, more than 1500 instruments were operating in the free world, in widely diverse fields. The present applications of the SEM are very broad (Kimoto and Russ, 1967; Wells, 1972), ranging from biology to geology, materials science (Russ, 1970), and semiconductor technology.

1 Instrumentation

Figure 1 shows in schematic block-diagram form the general construction and major components of the SEM. Actually, since the available instruments differ significantly in some design aspects, there are a number of choices possible in nearly every major component. A vacuum system is required for all electron beam instruments, since the scattering and absorption of electrons in air must be avoided. The vacuum systems used commercially include conventional oil diffusion pumps, with high speed and good reliability, as well as systems designed to produce a much cleaner vacuum, e.g., sublimation–ion pump combinations. The trend seems to be strongly toward the cleaner vacuums both to reduce contamination of the specimen and also to permit the use of brighter electron guns.

The source of electrons in the early instruments was a heated tungsten wire. However, the maximum current that can be made to strike the specimen depends strongly on the gun brightness, as will be discussed, and so there is interest in using brighter guns. One of the types now coming into widespread use is the heated lanthanum boride. Even higher in brightness, but as yet less widely used, is the cold field emission type. The increase in beam current over a tungsten wire can be as much as 6 to 10 times greater with the lanthanum boride, and several hundred times with the field emission tip at typical operating conditions. For most of the analytical purposes to be discussed, the data presented are based on conventional tungsten filaments, and the use of more intense beams will reduce the time needed to obtain the data proportionately.

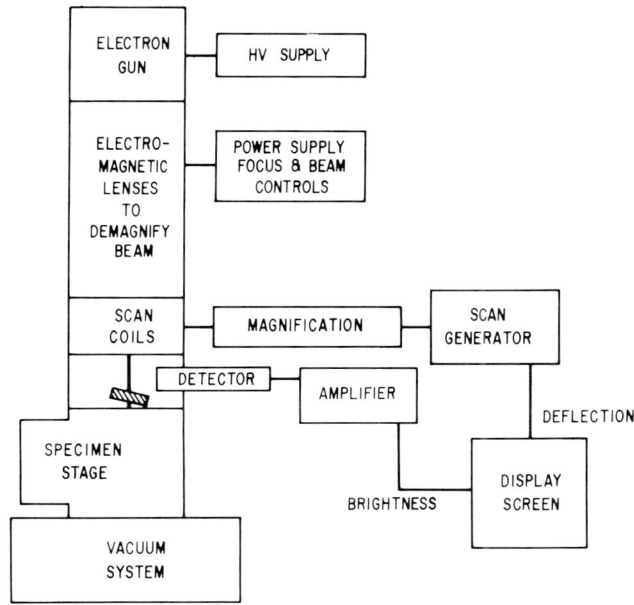

FIG. 1. Block diagram of scanning electron microscope.

The brightness of the gun and the source size also determine the number of lenses actually needed to achieve suitable demagnification of the electron beam. The number of lenses used in experiments has ranged up to at least four, but most conventional SEMs have three. By suitable gun design this can be reduced to two, one, or even none. Examples of each of these are performing well in routine use. It makes little difference (save in the manufacturing expense and complexity of operation) how many are used, so long as the beam can be reduced to a small enough diameter at the specimen surface. There are some clever designs of lens configurations that use the lens action to assist in the beam deflection for scanning, which may permit easy conversion to some diffraction experiments that will be discussed, but there is no theoretically optimum number of lenses.

The scanning facilities also vary widely, some using single deflection of the beam and others using two (or more) coils to deflect the beam, often so that it can be made to strike a single point in the specimen and rock, or vary its angle of incidence for diffraction studies. There is also a choice of continuous (or "ramp") scan generators which are generally more reliable, or step scan generators (or ladder circuits) that facilitate computer interfacing. And most present day instruments offer a TV-rate scanning capability for real-time examination of specimen, both to locate areas of interest and for use with heating, cooling, or mechanical loading devices.

Display facilities range from small long-persistence (slowly scanned) cathode ray tubes to conventional TV monitors. Controls, while they appear to differ greatly from machine to machine (especially in nomenclature), generally provide most of the necessary functions for signal selection and conditioning. One area still undergoing considerable development is the specimen facility. Provisions to accommodate large specimens and to move them in at least two horizontal directions and tilt are a minimum necessity. Beyond that more complex motions may be provided, and they may be made eucentric so that the field of view does not change during tilt and rotation. Little work has yet been done on the obvious desirability for heating and cooling facilities, rapid specimen exchange with airlock facilities, and externally controlled manipulators. Doubtless these will become available.

2 Principles

The purpose of all this expensive gadgetry is to take advantage of the various phenomena that occur when electrons strike material. At each point of the specimen surface, the signals shown in Fig. 2 are generated. Since the fine beam, typically under 10 mm (100 Å) in diameter, strikes only one point at a time, it is scanned over the surface to build up an area

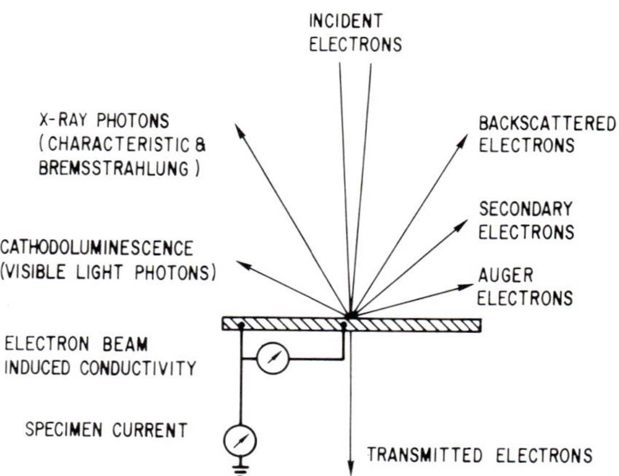

FIG. 2. Signals generated by the incidence of an electron beam on the specimen.

representation of the sample. The display tube is scanned in synchronization with the beam so that each point on the tube corresponds to a point on the specimen, and then the intensity of the spot in the tube is modulated in proportion to one or another of the signals shown. It is not at all obvious so far that the result will be useful, but in fact it is as we will see.

The range of magnification of the SEM is limited at the low end by the ability of the scanning circuitry and coils to linearly deflect the electron beam, and can be as low as 1 or 2×. Since the magnification is simply the ratio of the display tube dimension to the distance the beam is scanned on the specimen, the magnification can be increased by attenuating the signal reaching the scanning coils without affecting any other parameter such as focus. It is thus possible to go directly from very low to high magnifications on features of interest. The upper limit to magnification is the image resolution of the instrument. This will be discussed in detail for each of the signals, but in the most favorable case of secondary electrons (which are commonly used for the "pretty pictures" on magazine covers), it can be as high as 50,000×.

2.1 Secondary Electrons

As the incident electrons enter the specimen, they gradually lose their energy by collision with atoms in the sample, knocking out orbital electrons. These are called secondary electrons, and they are generated throughout the capture volume, shown schematically in Fig. 3. However, the secondary electrons have very low energy (less than 50 eV) and so they travel only a short distance before being thermalized. Only those electrons very near the surface can escape to depths ranging from a few angstroms in metals to a few hundred angstroms in insulators. Hence, they come primarily from the region immediately at the impact point of the beam, and carry information (we have not yet said what kind) about a region only a few tens of angstroms larger than the beam itself. This is why the image resolution is best with secondary electrons, with pictures showing considerably better than 10 mm (100 Å) now common.

The contrast in the secondary electron image comes from several possible variations in the specimen. One is the surface topography. If the surface angle relative to the beam changes by even a few degrees, there is an appreciable change in the number of secondary electrons that can escape, because more of the capture volume is near the surface. By the same argument, a sharp edge on the specimen will expose a great deal more area and thus emit a great number of secondary electrons. The result is that as the beam scans the specimen, the brightness of the cathode ray display

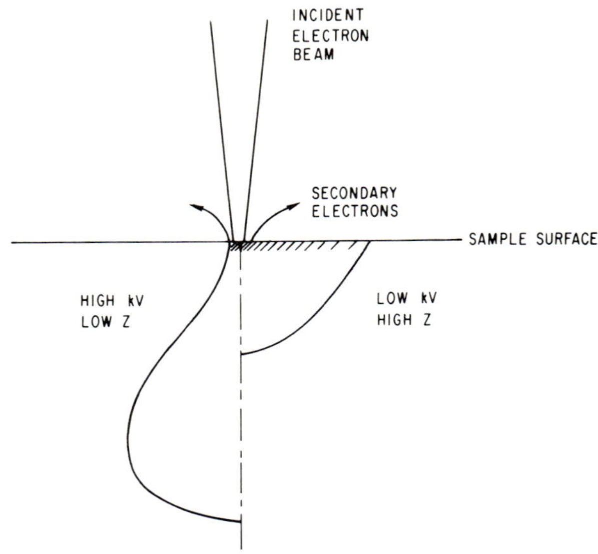

FIG. 3. Shape of electron capture volume varies with kV of incident electrons and specimen average atomic number. Effective volume of secondary electron emission is confined to surface layer.

tube changes gradually as the specimen slope varies and is very bright at edges.

The first effect produces contrast on rough specimens that is quite analogous to the normal appearance of objects viewed by eye in normal diffuse lighting, where the brightness varies gradually with surface angle. The enhancement of edges is not directly paralleled in visual experience, but the eye–brain combination tends to characterize images by edges anyway (for instance, consider engineering line drawings, sketches, and cartoons), and so this factor also adds to the ready interpretability. The secondary electron image of most rough surfaces is easily understood by the operator, and little interpretation is needed for qualitative work. This is perhaps a drawback since it leads the user to neglect the theory of the instrument and he tends often to oversimplify other, less obvious images. It also has led to a great deal of superficial work characterized by pretty pictures, and relatively little utilization of the ultimate capabilities of the SEM.

The secondary electron detector is generally composed of a grid held at some high positive voltage (e.g., 10 kV) in front of a scintillator or phosphor. The electrons from the sample are attracted and accelerated by the field of the grid, traveling in curved trajectories, and thus coming from regions on rough surfaces that are obscured to the line of sight from the detector. They strike the scintillator and the resulting light is conducted through a light pipe to a photomultiplier tube. The amplified signal can then be dealt with by conventional electronics and amplified, shaped by nonlinear circuits, or otherwise prepared for display.

For rough surfaces, the interpretation of the secondary electron image to quantitatively measure the relief usually requires the use of stereo pairs (Boyde, 1970). By taking two or more sequential pictures with the specimen tilted by a known amount (typically 5–10°), it is possible to use the parallax to measure elevation differences. The simplified formula for the case of tilting equal amounts about the horizontal and the complete equation for the more general case are shown in Fig. 4. For reasonably flat specimens where large angles of tilt (up to 15°) can be used, it is possible to achieve vertical resolution of several hundred angstroms, several times the resolution in the horizontal direction.

The secondary electron emission also varies with other factors such as specimen composition or applied electrical potentials. In general, the higher the work function of the surface the greater the emission, but virtually no work has been done to prove this quantitatively because of the difficulty in accurately measuring the secondary electron signal.

Depending on the detector to specimen geometry, it is usually possible to show contrast in the secondary electron image due to the application of low voltages to parts of the specimen (Oatley, 1969). Some efforts have been made to measure voltages in this way (Wells, 1969), but primarily

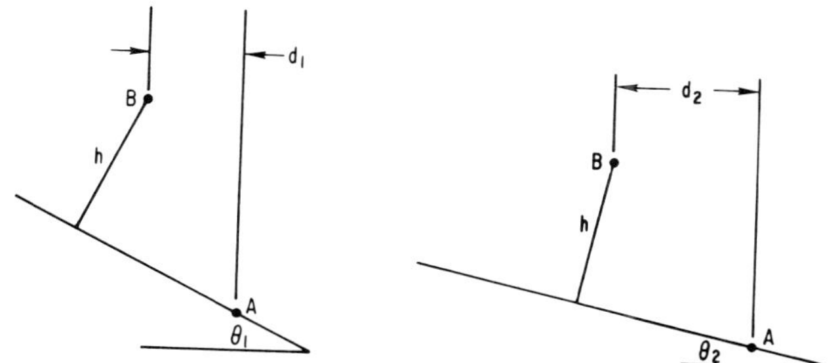

FIG. 4. Calculation of height of features by parallax measurement on stereo pairs.

$$h = \frac{d_2 \sec \theta_2 - d_1 \sec \theta_1}{\tan \theta_1 - \tan \theta_2}$$

If $\theta_1 = -\theta_2 = 4/2$ for equal tilt about the psi horizontal and D is the parallax $(d_2 - d_1)$ measured at magnification M, then $h = D/2M \sin(\pi/2)$.

the technique has been applied to locating failures in electronic components.

2.2 Backscattered Electrons

Some of the high-energy incident electrons (accelerating voltages up to 50 kV are commercially available) undergo Rutherford scattering from the sample atoms and reemerge from the surface. A suitable detector, often simply a Schottky-barrier photocell, can be used to detect these electrons, which still have very nearly the full original energy. The resulting image is in some ways like the secondary electron image, but there are a number of important differences. First, the backscattered electrons come from a greater depth in the sample, and because of the spreading of the electrons in the sample they represent a larger area, and thus give poorer resolution, rarely better than 200 nm (2000 Å).

Also, since the backscattered electrons come from deeper in the sample, they contain less information about the surface and more about the bulk material. And since they are of greater energy, they are not affected significantly by applied voltages or valence bonding in the material. The high energy means that they travel in essentially straight lines, and so there are completely dark shadow areas with typical rough surfaces. The image is still presented to the user as though he were looking down on the specimen from the top of the electron column, but instead of diffuse lighting, the effect is one of the sharply collimated light coming from the position of the detector.

The backscattered electron signal is monotonically dependent on the specimen's average atomic number, but with the fraction of electrons backscattered increasing as Z increases. This effect has in fact been used with both SEMs and microprobes to give elemental analysis in some cases. For flat samples, either with a naturally flat surface or one prepared by grinding and polishing the backscattered electron image shows contrast that is very useful for determining the distribution of phases of different composition as will be discussed in detail.

Another cause of contrast in the backscattered electron image that has barely been used to date but holds great promise is the so-called channeling pattern, or pseudo-Kikuchi pattern. This is contrast that arises because of internal diffraction of the incident electrons (Coates, 1969; Hirsch and Humphries, 1970). When the angle between the incident beam and the lattice planes in the sample permits Bragg diffraction to take place, the incident electrons penetrate more deeply into the specimen and hence fewer electrons backscatter to the detector. Since the angle of the beam varies by several degrees as it scans at low magnification, the result for a beam scanning over a perfect single crystal is an image that instead of being uniformly gray has dark lines running across it that can be used to index the crystal structure and orientation. (Because the interpretation of the patterns to obtain this information is identical to that for Kikuchi patterns in transmission electron microscopy, they are sometimes called pseudo-Kikuchi patterns.)

A more intriguing use of these channeling patterns than simply looking at large single crystals is the possibility of obtaining orientation information on single, tiny grains in a material (Joy *et al.*, 1971; Van Essen and Schulson, 1969). Since all that is required is to vary the angle of incidence of the electron beam with respect to the sample, there are several ways this could be done. Early attempts to fix the beam position and then rock the sample were hindered by the low speed and mechanical slope in stages. Electronic deflection of the incident beam has proved more satisfactory, although the combinations of deflection coils and lens action used to produce the effects vary widely. The possibility of obtaining channeling patterns from areas as small as 10 μm has been shown. This is larger than the backscattered electron resolution mentioned above because for this application the beam must be defocused so that the incident electrons are as nearly parallel as possible. In this case, the display tube image with its dark lines does not represent the specimen surface, but rather is a plane in reciprocal lattice space, and represents a single location on the specimen surface.

2.3 Auger Electrons

Another type of information is carried by the Auger electrons that leave the specimen. Normally included as a subset of the secondary electrons, the Auger electrons are those electrons from orbital positions in surface or near-surface atoms that were emitted by internal capture within an atom of an x ray produced by decay of an ionization in an inner shell. In other words, the incident electron produces an ionization by knocking out a bound electron (to become a secondary electron), and instead of the decay producing an x ray (which will be considered shortly), the photon is absorbed to produce a photoelectron from another shell. This rather complex process has a high probability of occurring for the very light elements, and the energy of the Auger electron can be used to determine the atom from which it came. Hence, the detection and measurement of these very-low-energy electrons can provide elemental analysis of the surface layers of the specimen (Zeitler, 1971). Auger electrons can only be meaningfully measured from the top few or fractional atomic layers because deeper ones lose energy and blend into the background. However, for the analysis of light elements (e.g., O, C) on surfaces with which they are reacting, the technique has great potential, as shown by work done with Auger electrons in LEED experiments.

Several difficulties hamper the ready application of the technique to the SEM. The first is obviously the poor vacuum present in many commercial instruments. The contamination of the specimen surface with carbon under the action of the electron beam will prevent measuring any other element of interest after a very short time. Also, the high percentage yield of Auger electrons for light elements as compared to x rays is more than offset by the fact that few are produced near enough to the surface to escape and so the total intensity is extremely low. Finally, it is difficult to design an electron spectrometer to measure these electrons with the necessary accuracy and also to collect a large fraction of them without taking up so much space near the specimen that other functions of the SEM and other signal detectors are compromised.

2.4 Specimen Current

By measuring the fraction of the incident beam that remains in the sample, one can obtain the complement of the backscattered electron signal. But since it is unusual for the backscattered electron detector to cover a large solid angle with respect to the specimen, so that it becomes strongly dependent on shadowing effects, the specimen current image is sometimes useful to sort out chemical or orientation effects from topographic effects. The resolution of the image is poor—$\frac{1}{2}$ to 1 μm typically—because of spreading in the sample.

In addition to this use, its main purpose is to provide a ready measurement of beam current when that is needed to quantify some other measurement that depends on the beam intensity. For this purpose, it is gradually becoming common to provide a small Faraday cage in the corner of the specimen holder so that the measurement of the total beam current can be readily obtained when needed. The use of the specimen current instead of the true total beam current will cause major errors because the backscattered fraction is large (e.g., $\sim$30% for Fe) and dependent on composition.

2.5 Transmitted Electrons

If the specimen is reasonably thin (a purposely loose term that will be refined), some of the incident electrons will go through and can be detected by another scintillator-photomultiplier tube-type detector or some other type, as we will see. If one uses a specimen like that normally encountered in conventional transmission electron microscopy, the resulting image is quite similar in overall appearance (MacDonald, 1971). The resolution is considerably poorer in most SEMs than that which is now routine in TEMs because few commercial SEMs are designed to produce an incident beam smaller than about 100 Å, and so it is not possible to resolve detail smaller than that with the scanning transmission image.

Even so, the image is usually good enough to identify major features that are known from TEM studies, so that the electron beam can be located and positioned for some other experiment or analysis. Furthermore, if the electron detector has the ability to separate the inelastically scattered and elastically scattered electrons, the contrast in the resulting image can be improved markedly. This can be easily done by using an annular aperture or two detectors with one hooked to an annular scintillator and light pipe. In either case, the contrast with inherently low-contrast materials, such as unstained biological tissue, is often many times greater than with the TEM, and permits eliminating the stain to simplify further analysis.

Also, if an electron spectrometer is used to measure the energy loss of the transmitted electrons, it is possible to perform some types of analysis on features in the specimen. This technique has been refined by A. V. Crewe and his co-workers at the University of Chicago (1970), who have constructed a scanning transmission electron microscope with a high-intensity field emission gun and with an electron spectrometer under the specimen. By using the spectrometer to separate the electrons that have lost energy from those that have not, and then electronically ratioing the inelastically to elastically scattered electrons, they have obtained an image signal that is dependent only on atomic number and is sensitive enough to record the presence of single heavy atoms on a substrate. The work is continuing toward the identification and localization of atoms in biologically interesting molecules.

Another use of the transmitted electron image is for specimens thicker than those normally used for TEM studies. In the conventional TEM, as specimen thickness increases, the energy loss in the electrons that penetrate the sample increases, so that they are no longer monoenergetic: then the chromatic aberrations that are inherent in electromagnetic lenses make it impossible to focus the electron image sharply, and the result is a very rapid loss of resolution with thickness (Shibatomi *et al.*, 1971). This has been the major reason behind the development of high 1-1 and ultrahigh-voltage TEMs, now up to 3 MeV.

In the SEM, however, the image resolution is controlled solely by the incident beam size, which is a function of lenses above the specimen where the electrons are monoenergetic. There are no lenses below the specimen; hence, the resolution can be maintained up to rather great specimen thicknesses so long as the electron energy is sufficient for penetration. Actually, there is a slight drop-off due to multiple scattering events within the specimen, but this is practically negligible compared to the situation with the TEM. It is usually possible to examine specimens with scanning transmission microscopy several times thicker than could be imaged with a TEM of the same accelerating voltage. For example, adequate images of 0.2 to 30 nm (2 to 300 Å) resolution can be obtained at 50 kV with a 10- to 15-nm (100- to 150-Å) incident beam on 1- to 2-μm-thick embedded and sectional biological tissue. Also, there are now some specially designed SEMs, such as the one built by Crewe, that offer the promise of resolution competitive with the TEM (incident electron beam diameters of a few angstroms).

2.6 Electron-Beam-Induced Conductivity

When high-energy incident electrons strike semiconductor material, another effect can be observed. Since the process of electron deceleration involves the creation of free electrons and holes, the production of these free-charge carriers can produce a current across an intrinsic junction. The qualitative uses of this phenomenon are chiefly to locate the junctions by scanning the electron beam over the sample and imaging the current flowing in the device (which must of course be connected to wires leading outside the vacuum system to a suitable power supply). Then bright lines on the image will identify the junctions. This image is often mixed electronically or photographically with the secondary electron image to facilitate interpretation by showing the physical features of the specimen surface.

Because this technique involves the capture of the incident electrons well below the specimen surface, and that depth can be controlled by the electron accelerating voltage used, this method can be used in addition to study the junctions in depth. In order to quantify this measurement, it is necessary to know the distribution of energy deposition in the sample (we will also need this for the x-ray quantitative analysis to be discussed). There are several mathematical ways to evaluate this; considerable data are available on the interactions of electrons with matter (Wittry, 1970).

Much of this work borders on the electronic applications of the SEM (Thornton, 1968), which also include using it as a fabricating tool to expose photoresist in the production of integrated circuits. We will not discuss this application in detail since it is not directly related to the analytical uses of the SEM.

2.7 Cathodoluminescence

Some materials emit photons of visible or near visible light when excited by electrons. Examples are many minerals, some organic compounds, and some industrially important chemicals. The light can be easily detected with a photomultiplier tube, but this gives little information about wavelength. Dispersive methods for measuring the wavelength are too inefficient to give useful results, especially with conventional SEMs having relatively low beam currents (e.g., with heated tungsten filament guns) and so have been used very little. Even filters placed in front of the photomultiplier tube reduce the intensity so much that they cannot be used in many cases. This has made the quantitative interpretation of cathodoluminescence results difficult or impossible.

Nevertheless, the sensitivity of the technique can be very great. Changes in composition at the ppm level and below can significantly alter the color or intensity of the emitted light. These low concentrations in such small volumes are far below detectability with other methods. It is not usually easy to interpret the meaning of the changes quantitatively, but they still give information about subtle variations in the sample. No comprehensive data on cathodoluminescence intensity and color from various materials is yet available, but Task Group E.04.15.03 of the American Society for Testing and Materials is compiling data as they become available, and eventually reference tables may exist. They would make it possible to infer, for instance, that a change from blue toward yellow in some mineral compound was associated with an increase in trace Fe content. The availability of higher intensity electron beams from SEMs in the future will permit more sophisticated wavelength determination as well.

Since the light photons are generated throughout the electron capture volume, and since most Cl-emitting materials are nonconductors and hence fairly transparent to their own radiation, the detector signal comes from a large region. The size is very dependent on incident electron voltage and for some particular applications can be reduced to 100 nm (1000 Å), but in most typical cases it is $\frac{1}{2}$ to 1 μm.

2.8 X-Ray Analysis

The ionization of atoms in the specimen by the incident high-energy electron results in the emission of characteristic x rays from the elements present. In addition, the deceleration of the electrons, since they are charged particles, causes the emission of continuous or Bremsstrahlung radiation. The total flux of x rays coming from the specimen is therefore a combination of a continuous energy distribution (up to the maximum energy of the incident electrons), and a series of discrete energies corresponding to the energy level transitions in the atoms which can be used to unequivocably identify the atoms present.

Detection of the x rays can be accomplished by any of several means, including a simple Geiger tube. By simply detecting and displaying the entire x ray signal, an image can be presented that contains some information about the specimen. Since x-ray photons travel in straight lines, there will be shadowing effects from topographic features, and in addition variations in image brightness from regions of one composition to the next. In general, the flux of x rays decreases as the average atomic number decreases because of the reduced fluorescence yield (and rise in Auger electron production noted before), and the x-ray flux also decreases as the absorption edge energy or energy required to excite the atom by knocking out an orbital electron increases toward the incident electron energy. Since this ionization energy rises as atomic number rises, the result is contrast in the x-ray image that ranges from brightest for some intermediate atomic number that depends on the accelerating voltage to less bright for both higher and lower atomic numbers.

Because of the low total efficiency of producing x rays as compared to the various electron signals, and because of the difficulty in interpreting the image, the use of the total x-ray flux to produce an image has been very rare. On the other hand, the measurement of the x-ray energy to obtain analytical data is a very common technique. The classical method of determining x-ray energy, as exemplified by the electron probe microanalyzer (see Chapter 6), is to take advantage of the wave nature of the photon and measure the wavelength. The wavelength is given by $\lambda = hc/E$ where E is the h is Planck's constant, and c is the speed of light. The phenomenon of diffraction can be used to measure the wavelength by using a crystal of known atomic structure (or actually several of them to cover the elements of interest) and a mechanism to control the angle at which the x rays strike the crystal. Then the Bragg equation predicts that diffraction will occur when the angle θ is related to the wavelength by $n\lambda = 2d \sin \theta$, where n is any integer, λ is the wavelength, d is the atomic spacing in the crystal, and θ is the angle between the crystal and the entering and leaving x rays. More complete details on the design of these spectrometers can be found in the chapter on electron probe microanalysis and the chapter on conventional x-ray fluorescence analysis.

These types of spectrometers have found very little application to the SEM because of their low efficiency (Sutfin and Ogilvie, 1970). The exacting geometrical restrictions imposed by diffraction make it difficult for the spectrometer to accept a very large fraction, or solid angle, of the x rays emitted in all directions by the sample. Furthermore, 80–90% of the x rays are lost in the diffraction process itself, so that the count rate at the detector (usually a gas proportional counter) is low. Since the electron beam currents in the SEM are very low, typically four to five orders of magnitude below those on the electron probe microanalyzer, these so-called wavelength-dispersive analyzers are unsuitable for general-purpose use with the SEM.

Since this is an important point, let us consider the diameter of the electron beam in the SEM: $d = (2C^{1/4}/\pi^{3/4})(i/\beta)^{3/8}$ where C is the special aberration constant of the lens, i is the beam current, β is the gun brightness $[S(eV/kT)]$, where V is the accelerating voltage, T is the temperature, and S the electron emissivity. In order to reduce the diameter from the size typical in the microprobe (between $\frac{1}{2}$ and several micrometers at the high beam currents used for analysis) to the range of less than 10 nm (100 Å) desired for the SEM, the beam current must be reduced from about 10^{-7} to only 10^{-11}–10^{-12} A. This drop in intensity would reduce the count rate for a wavelength dispersive spectrometer from a comfortable 10^5 counts per second to only about 1 count per second.

Therefore, most SEM users have turned to a new type of x-ray spectrometer for analytical data (Russ, 1971a). This is the more recently developed technique of energy-dispersive x-ray analysis (EDXA), so called because instead of using diffraction to measure the x-ray wavelength, it directly measures energy and presents the data as a spectrum of intensity versus energy. Since EDXA is a relatively new technique, a brief discussion on how the analyzer works may help the reader.

The x rays enter a detector consisting of a pure wafer of silicon, as shown schematically in Fig. 5. This silicon acts as a solid-state ionization chamber, and the incident x rays lose energy by the creation of photoelectrons and

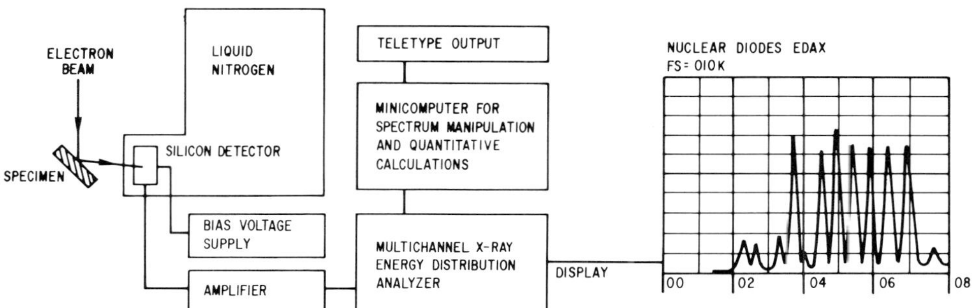

FIG. 5. Block diagram of energy dispersive x-ray analysis system for SEM.

the subsequent ionization of silicon atoms. Since each ionization requires 3.8 eV of energy, the number of atoms ionized and the number of electrons freed depend linearly on the energy of the incident x ray. The total charge deposited in the silicon is collected by an applied bias voltage and integrated electronically to produce a small pulse. Since all of these operations are linear, the pulse height is a measure of the x-ray energy, and since the processes are all very fast, each x ray is measured as it enters the detector. The detector and preamplifier are cooled with liquid nitrogen to reduce electronic noise but have no moving parts to require adjustment or maintenance.

The pulses coming from the detector are measured by a special-purpose computer called a multichannel analyzer. Each pulse charges up a small capacitor, which is then discharged. The length of time required for the voltage to drop is measured by a high-frequency clock, and the time is used as an address to store the count for that x ray in a spectrum. This spectrum is normally presented for the operator on a cathode ray tube or TV display as shown in Fig. 5. The spectrum shows the distribution of energies of the x rays from the specimen, both the Bremsstrahlung and the peaks corresponding to the characteristic energies from the elements present. Further data processing may also be provided, as shown in the figure, to reduce the data to quantitative analysis of the spectrum. Small on-line minicomputers are available with software to perform the operations that will be described below.

It is beyond the scope of this chapter to discuss in detail the instrumental parameters that affect the performance of EDX analyzers. In general, the following are advantages of EDXA over wavelength-dispersive analysis:

1. The silicon detector can be placed very close to the specimen to collect a large solid angle of the emitted x rays and is virtually 100% efficient. This provides adequate count rates for most types of specimens at normal SEM operating conditions.

2. The lack of geometrical focusing requirements permits operation at low magnifications and with rough surfaces. These are normal for the SEM but cannot be used in microprobes because x rays that originate at points off the focusing (or Rowland) circle of the spectrometer will not be detected, giving rise to apparent variations in composition that are not truly present.

3. X rays of all energies are detected simultaneously, so that a complete spectrum is obtained at once. With wavelength dispersion only a single wavelength can be detected at a time and even with the high count rates present in the microprobe, and with several spectrometers equipped

with several appropriate crystals for each wavelength range, it commonly takes $\frac{1}{2}$ to 1 hr to completely analyze an unknown sample. The EDX Analyzer can detect all major elements within seconds, and provide sufficient data for quantitative analysis of major and minor elements in minutes, with no danger of missing an unexpected element.

4. The EDX Analyzer has no moving parts or mechanical adjustments, and requires very little training or knowledge to use. This is particularly important for the typical SEM user, who is primarily concerned with his specimens and their examination, and only secondarily in the analytical results. The spectral peaks are readily identified with elements, and there are no "high-order" diffraction lines to be wary of.

On the other hand, the conventional wavelength-dispersive analyzer has some areas of advantage too.

1. Light-element detection for elements down to boron in the periodic table is well established with crystal diffraction, although the ability to obtain quantitative results is still quite poor because of poor data on absorption coefficients, electron stopping power, and other fundamental parameters. The EDX detector is normally sealed in its own vacuum system by a thin beryllium window that absorbs the low-energy x rays from B, C, N, and O. Detection of these elements has been accomplished with special detectors having no window, but at the time of writing these appear to be several years away from commercial availability (Russ, 1971b).

2. The spectral resolution (width of peaks) is much better for a wavelength-dispersive spectrometer, typically in the 10–20 eV range as opposed to the 150–200 eV range of the EDX analyzer. The principal peaks from adjacent elements are well separated and present no difficulties, but there are some interferences between peaks from different x-ray emission lines from different elements, such as the Cr K_β and the Mn K_α, or the S K and the Pb M lines, that cannot be separated and will not be even if EDXA performance inproves to the theoretical limits. For these cases, computer techniques are used to strip the peaks apart (Russ, 1971d). (See Fig. 6.)

2.8.1 *Experimental Procedures and Applications*

X-ray detection can be used on the SEM to present information on specimen composition in several forms. One is roughly analogous to the other images we have described. The spectrometer is set to a particular element of interest and an image produced by the signal detected. Because the production of photons is rather low in intensity, the result is not a continuous gray scale as for the electron images. Instead, the photons of energy corresponding to the element of interest (as well as those Brems-

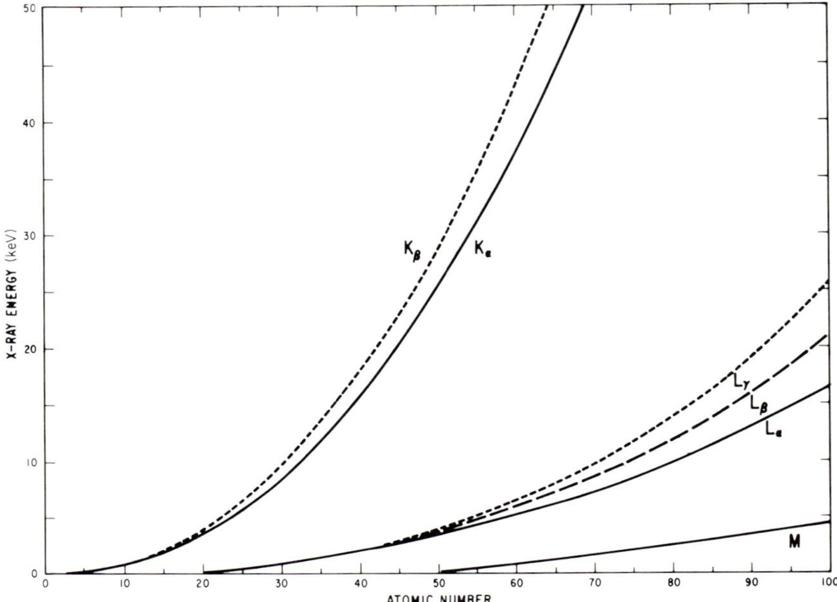

FIG. 6. Energy of important x-ray emission lines as a function of atomic number.

strahlung x rays of the same energy) are used to produce a bright dot on the display CRT at a point corresponding to the position of the electron beam when the x ray was detected. The resulting image shows clusters of these dots in regions where the element is high in concentration, and vice versa. By comparison of this image to the secondary electron (or other) image, or by photographically or electronically superimposing them, it is possible to identify the regions of the sample containing concentrations of the element. Because of the highly statistical nature of the dot image, it is difficult to make even the most qualitative judgments of relative concentration. Also, it is difficult to show concentration differences smaller than several times, and very low amounts of the element of interest may not be visible above the natural Bremsstrahlung background; this is especially true with EDXA compared to wavelength dispersive spectrometers as used on the microprobe because of the poorer peak to background ratios.

A source of potential confusion in interpreting x-ray images with the SEM lies in the nature of the samples themselves. Most SEM specimens are rough and the x-ray intensity of both Bremsstrahlung and characteristic radiation varies from point to point because of absorption effects.

This makes it very desirable to examine x-ray images made with both the characteristic x rays of the element of interest and with Bremsstrahlung x rays of a similar energy. Differences between these two images reveal the effects of composition as distinct from topography.

When flat samples are used, it is practical to use x-ray images in combination with electron images to obtain quantitative information on the size and distribution of phases in the specimen (Braggins *et al.*, 1971). This is called quantitative metallography, modal analysis, or stereology, in various fields of application. The techniques widely developed for light microscopy and the various commercially available instrumentations can be utilized with the SEM. It is especially desirable to use combinations of signals for this (Dorfler and Russ, 1970), since the x-ray signal (and in some cases the cathodoluminescence signal) can be used to identify the phase, and the secondary or backscattered electron image with superior resolution and gray scale can be used to delineate the phase boundaries. The technique is useful for polished sections through materials or for particles on a substrate, such as filters with air pollutants. The parameters that can be measured automatically are phase (or particle) size and size distribution, and shape and orientation. Values for the volume fraction of a phase, the mean free path, the degree of anisotropy, the presence of banding or striations, and other similar measurements that have been shown to relate to material properties can be directly computed.

Because of the difficulty in determining relative composition from x-ray distribution images, another common method of presenting information is the line scan. In this case the beam is slowly scanned along a line across the sample, generally a line selected to cross a feature of interest such as a phase boundary, diffusion zone, etc., while the intensity of x-ray signal for a given element of interest is monitored. Frequently this is done with a rate meter and the count rate displayed as a function of position directly on the display CRT of the SEM so that it can be superimposed on the secondary or backscattered electron image. Since the beam is scanned slowly, more x rays are counted for each point on the specimen than with the x-ray dot image. This gives adequate statistics to permit fairly sensitive determination of concentration variations along the line. Variations of as little as 1% relative concentration can be detected when suitable conditions and relatively long times such as several hundred seconds per line are used.

When true rather than relative concentrations are desired, it is necessary to use point analysis. This is also used when the greatest sensitivity is required or when all the elements present in a particular feature of interest are to be determined. For this application, the beam is not scanned but is positioned with reference to the image on the display tube and held stationary for the time necessary to obtain the analysis. In many cases this requires as little as seconds for the major elements and perhaps 100–200 sec for the minor elements. Concentrations below 0.1 to 0.01% are difficult to detect because of the Bremsstrahlung background. The resulting spectrum can be examined on the display of the EDXA system or recorded in a variety of ways, either analog or digital. To obtain true or quantitative analysis in general requires further processing of the raw data. The corrections for the absorption and secondary fluorescence of x rays within the sample are the same as for the conventional microprobe. The corrections can be made mathematically provided that proper subtraction of the Bremsstrahlung background is made first. This is not a simple operation because of the presence of absorption edges in the spectrum, but can be accomplished with a small on-line minicomputer (Russ, 1971d).

Another common method for obtaining quantitative analysis is by comparison to known standards of similar composition. Calibration curves or their mathematical representation can be used to convert the measured intensity to true percentages. The accuracy of this technique can be better than with the purely mathematical methods, and is limited primarily by the quality of the standards available. Relative accuracy of 1 to 2% is not difficult to achieve with appropriate standards, while 2 to 5% relative accuracy can usually be obtained using the theoretical corrections for absorption and fluorescence. Poorer results are usually found for trace elements because of the contribution of the Bremsstrahlung background and the difficulty of accurately removing it. Comparative figures for the wavelength-dispersive spectrometers used with microprobes can be found in the chapter on that instrument.

A very different type of point analysis is becoming more widely used with thin sections, especially ones of biological tissue. The high level of spatial resolution and sensitivity, along with quantification of the amount of the element present, promises to open important new areas of research and thus warrants some further discussion.

The size and shape of the volume emitting x rays is very different in a thin section than in the bulk material previously illustrated because most of the electrons exit from the bottom before much scattering can occur (Russ, 1971e). There is still some lateral spread, which we have calculated can be well approximated as a truncated cone with a top diameter equal to the incident beam diameter and a cone half-angle that for 50-kV electrons and a specimen composed primarily of plastic embedding medium is equal to 40° to include 90% of the generated x rays, and 18° to include 50% of the signal, as shown in Fig. 7.

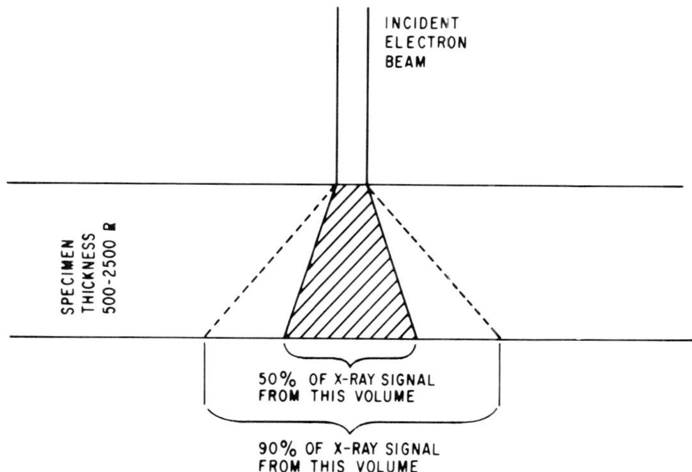

FIG. 7. Volume of x-ray emission in thin section of embedded biological tissue.

This means that by reducing the specimen thickness, very small areas can be analyzed [e.g., 50% of the signal from an area 30 nm (300 Å) in diameter at the midplane of a 50-nm (500-Å) section]. Such a small area contains very little material and so the signal is low. Most of the atoms present in organic material are too light to be detected, but it is useful to detect, analyze, and locate the trace amounts of heavier elements in the tissue. Sensitivities to less than 10^{-17} g are possible but require analyzing times of 15 to 30 min.

The use of slightly thicker sections containing more material improves the signal strength but at some cost in spatial resolution. Improvements in both sensitivity and localization can be obtained by eliminating the embedding plastic and water vapor through freeze-drying and sectioning, since the scattering angles and background will be reduced, and the heavy elements effectively concentrated.

3 Conclusions

Because the SEM is still a relatively new tool for the analytical laboratory, it has not yet been fully exploited. This is compounded by the intuitive value of the normally used secondary electron image, which provides so much new and valuable information in its own right that few users have yet begun to push the instrument to its ultimate capability for providing analytical and quantitative information about specimens.

Each of the various signals can be used to obtain some information of analytical use and value. Even more valuable data can be obtained by combining multiple signals. The use of energy-dispersive x-ray detection permits obtaining elemental analysis under typical SEM operating conditions, and with recent and continuing development is becoming very competitive with the wavelength-dispersive spectrometers used on conventional microprobes. Elemental analysis of microvolumes with sensitivities to less than 0.01% and quantitative accuracy of a few percent can be obtained.

The crystal structure and orientation of volumes as small as 100 μm^3 can be studied although with present instrumentation the techniques are cumbersome and time consuming, and the interpretation difficult.

Signal processing to measure the size and distribution of phases and particles using techniques and hardware similar to those employed in normal light microscopy is also feasible with the SEM, and can provide measurements of structural parameters that are important to the understanding of the physical properties of materials.

References

Boyde, A. (1970). Practical problems and methods in the three-dimensional analysis of scanning electron microscope images. *Proc. Annu. SEM Symp. IITRI*, **3**rd, p. 105.

Braggins, D. W., Gardner, G. M., and Gibtard, D. W. (1971). The Applications of image analysis techniques to scanning electron microscopy. *Proc. Annu. SEM Symp., IITRI*, **4**th, p. 393.

Coates, D. G. (1969). Pseudo-Kikuchi orientation analysis in the SEM. *Annu. SEM Symp., IITRI*, **2**nd, p. 27.

Crewe, A. V., Wall, J., and Langmore, J. (1970). Visibility of single atoms. *Science* **168**, 1338.

Dorfler, G., and Russ, J. C. (1970). A System for Stereometric analysis with the SEM. *Proc. Annu. SEM Symp., IITRI*, **3**rd, p. 65.

Hirsch, P. B., and Humphries, C. J. (1970). The Dynamical theory of SEM Channeling patterns. *Proc. Annu. SEM Symp., IITRI*, **3**rd, p. 449.

Joy, D. C., Booker, G. R., Fearon, E. O., and Bevis, M. (1971). Quantitative crystallographic orientation determinations of Microcrystals present in solid specimens using the SEM. *Proc. Annu. SEM Symp., IITRI*, **4**th, p. 497.

Kimoto, S., and Russ, J. C. (1967). The characteristics and applications of the SEM. *Amer. Sci.* **57**, 1, 112–133.

MacDonald, N. C. (1971). Auger electron spectroscopy for scanning electron microscopy. *Proc. Annu. SEM Symp., IITRI*, **4**th, p. 89.

Oatley, C. W. (1969). Isolation of potential contrast in the SEM. *J. Sci. Instrum. Ser.* **2**, **2**, 742.

Oatley, C. W., Nixon, R., and Pease, R. (1965). Scanning electron microscopy. *Advan. Electron. Electron Phys.* **21**.

Russ, J. C. (1970). Use of the SEM in the materials sciences. Spec. Tech. Publ. 480, p. 214. ASTM, Philadelphia, Pennsylvania.

Russ, J. C. (1971a). Energy dispersion X-ray analysis on the SEM. Spec. Tech. Publ. 485, p. 154. ASTM, Philadelphia, Pennsylvania.

Russ, J. C. (1971b). Light element analysis using the semiconductor X-ray energy spectrometer. Spec. Tech. Publ. 485, p. 217. ASTM, Philadelphia, Pennsylvania.

Russ, J. C. (1971c). Progress in the design and application of energy dispersive X-ray analyzers for the SEM. *Proc. Annu. SEM Symp., IITRI, 4th,* p. 65.

Russ, J. C. (1971d). The EDIT system for computer reduction of energy dispersive X-ray data. *Proc. Int. Conf. X-ray Opt,* **6**th, *Osaka.*

Russ, J. C. (1971e). Spatial resolution of X-ray analysis with solid and thin specimens. *Proc. Electron Microsc. Soc. Amer.* **29,** 54.

Shibatomi, K., Yamanoto, T., and Koike, H. (1971). Observability of very thick specimen by using transmission scanning electron microscope. *Proc. Electron Microsc. Soc. Amer.* **29,** 26.

Sutfin, L. V., and Ogilvie, R. E. (1970). A comparsion of X-ray analysis techniques available for scanning electron microscopes. *Ann. SEM Symp., IITRI,* **3**rd p. 17.

Thornton, P. R. (1968). "Scanning Electron Microscopy." Chapman and Hall, London.

Van Essen, C. G., and Schulson, E. M. (1969). Selected area channeling patterns in the SEM T. Mat'ls. *Science* **4,** 336.

Wells, O. C. (1972). No attempt is made in this survey paper to describe individual applications. Comprehensive bibliographies are available, the most broad-ranging of which is compiled by Wells, IBM Corp., Yorktown Heights, New York, and published each year in the *Proc. Annu. Scanning Electron Microsc. Symp.* IIT Res. Inst., Chicago, Illinois.

Wells, O. C. (1969). Method for measuring voltages in the SEM. *Annu. SEM Symp., IITRI,* **2**nd, p. 397.

Wittry, D. B. (1970). Electron beam interactions in solids. *Proc. Annu. SEM Symp., IITRI,* **3**rd, p. 409.

Zeitler, E. (1971). Scanning transmission electron Microscopy. *Proc. Annu. SEM Symp, IITRI,* **4**th, p. 89.

CHAPTER 17

Ultraviolet Photoelectron Spectrometry

John J. Uebbing*

Varian Associates
Palo Alto, California

Introduction

Photoelectron spectroscopy is well recognized as a uniquely powerful tool for determining the electronic structure of matter. In this technique, the sample is exposed to a flux of monochromatic photons and the resulting photoelectrons are energy analyzed in an electron spectrometer. In particular, vapor phase or molecular photoelectron spectroscopy using ultraviolet photon sources and high-resolution electron energy analyzers has brought forth a large amount of detailed information on the valence electron binding energies of free molecules. This has allowed a direct and relatively unambiguous experimental comparison with molecular orbital calculations.

The photons liberate electrons from the molecule by the process

$$M \xrightarrow{h\nu} M^+ + e$$

* Present address: Hewlett-Packard, Palo Alto, California.

M^+ may be the stable state of the ion, or if a deeper lying valence electron is removed, M^+ will be an excited state of the ion. The photoelectron spectrum is a faithful representation of the excitation energy of the ion. In the limit of Koopmans' (1934) approximation, this is the same as the energy of the orbital occupied by the electron. The ion will generally be left in an excited vibrational state and this will give rise to observable vibrational structure on the photoelectron spectrum. This vibrational structure can then be further analyzed to determine something about the nature of the orbital.

The most comprehensive review of this work is contained in the book, *Molecular Photoelectron Spectroscopy* (Turner *et al.*, 1970). Here is gathered a large number of spectra and their interpretation.

1 Instrumentation

Helium resonance radiation at 584 Å (21.21 eV) has proved to be the most useful photon source for this work. Since there are no convenient windows in this spectral range, the light flux is admitted to the sample chamber through a series of differentially pumped capillaries. According to Samson (1969), this source has a line width from 1 to 6 meV.

Neon resonance radiation (16.85 and 16.67 eV) is sometimes used because of its lower energy. Normally, the resonance lamps are operated at about 100 mTorr of gas pressure and 50 mA of current. By reducing the pressure to 10 mTorr and raising the discharge current to 500 mA, substantial amounts of radiation from He(II) (40.8 eV) and Ne(II) (26.9 eV) can be obtained.

The earliest instruments described (Al-Joboury and Turner, 1963) used a retarding grid type of electron spectrometer. However, deflection-type analyzers have shown markedly superior resolution and all of the commercial instruments are of this type. Varian, AEI, McPherson, and Vacuum Generators offer ultraviolet molecular photoelectron spectroscopy accessories for their x-ray photoelectron spectrometers. Perkin-Elmer offers an instrument that is designed solely for ultraviolet work. There are two methods of scanning the electron spectrum: One is by operating the electron analyzer at constant energy and sweeping a retarding voltage applied to the sample chamber; the other is to sweep the analyzer by varying the voltage across the deflection plates. The first has the advantage of maintaining constant resolution and high sensitivity over the entire energy range and the disadvantage of a large peak at zero electron energy and possible variation in sensitivity versus energy due to changing electron optics in the retarding lens. The second has the characteristic of a resolution proportional to electron energy. Figure 1 shows a cross-sectional view of the Varian

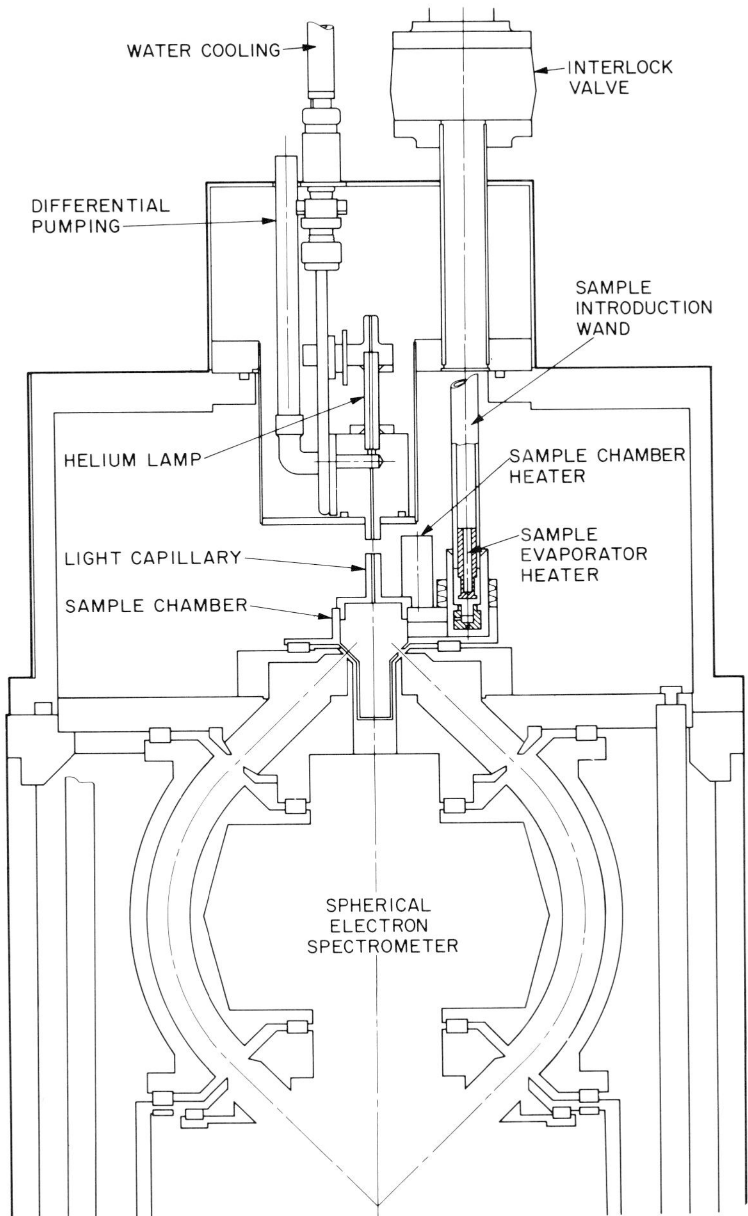

FIG. 1. Cross-sectional view of Varian photoelectron spectrometer equipped with ultraviolet accessory. The system is pumped by a turbomolecular pump from below.

instrument equipped with its ultraviolet accessory. This instrument operates by sweeping the sample potential. In addition to the ultraviolet lamp with its differential pumping, sample chamber, and the electron energy analyzer, provision must be made for sample introduction and vaporization as well as the admission of a calibration gas. Typical sample pressures run from 10 to 100 mTorr and the sample chamber can be heated to 250°C. The sample consumption rate is on the order of 10^{-2} Torr liters/sec or 20 mg/hr for a compound with a molecular weight of 100.

In principle, photoelectron spectroscopy is a nondestructive method of analysis. However, it is not generally practical to recover the sample from the output of the high-vacuum pump.

It is quite important for the attainment of high resolution that the analyzer surfaces have uniform work function (contact potential). It is even more important that the sample chamber have a uniform work function. A useful expedient in this regard is colloidal graphite, which is sprayed over the sensitive surfaces and allowed to dry. Experiments (Parker and Warren, 1962) indicate that this gives quite a uniform work function.

The photoelectron lines exhibit a broadening proportional to the molecular velocity and the electron velocity. This amounts to 21 meV for H_2 at 300°K with 584 Å illumination, whereas for argon it is only 4.6 meV. For light molecules such as H_2, rotational structure (Asbrink, 1970) is a further cause of line broadening, giving sidebands on the order of 30 meV lower in binding energy. The best resolution that is currently obtainable in commercial instruments is about 15 meV. This will resolve rotational structure only with great difficulty but is almost always sufficient to resolve the available vibrational structure.

There are small unavoidable shifts in observed binding energy caused by absorption of sample molecules on the sample chamber wall and positive ion space charge in the sample chamber, as well as the temperature dependence of physical dimensions and molecule absorption. For precision binding energy measurements, it is essential that a calibration gas be introduced into the sample chamber along with the sample. Gases used are argon, and argon–xenon mixtures. Often air or nitrogen will give an inadvertent and useful calibration. Lloyd (1970) lists a number of useful peaks and their binding energies.

2 Atoms

The photoelectron spectrum of argon is shown in Fig. 2. All spectra were taken on the Varian instrument. Both krypton and xenon show similar spectra. The two peaks correspond to the $^2P_{1/2}$ and $^2P_{3/2}$ states of the ion.

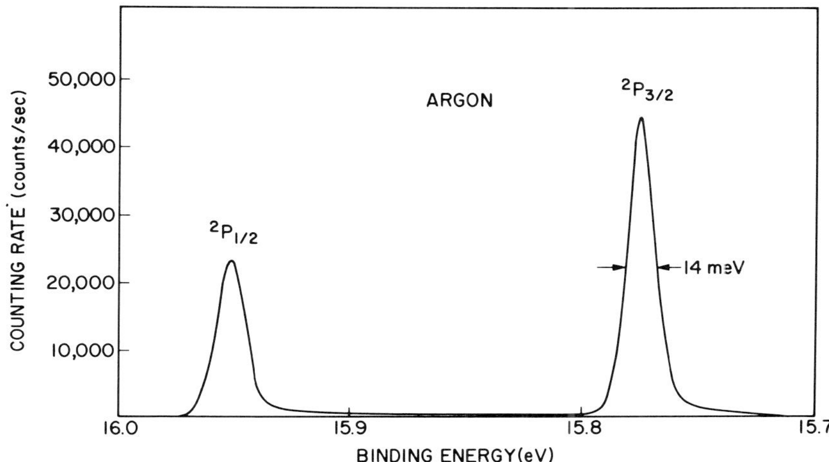

FIG. 2. Photoelectron spectrum of argon.

The amplitude ratio is not exactly 2:1 because the ionization cross section is not the same for the two states. These peaks are used for calibration since their energy levels are known accurately and the peaks are very narrow owing to the lack of rotational or vibrational broadening and minimal thermal motion broadening. In this most favorable case, the sensitivity of the technique is such that the instrument consumes 10^{13} argon atoms to produce 7 counts at the top of the argon peak.

3 Diatomic Molecules

The spectra of diatomic molecules illustrate some of the fundamental features of molecular photoelectron spectroscopy. One of the most important of these is how the vibrational structure in the different bands can be an aid in determining the nature of the orbitals involved. In a diatomic molecule, a particular orbital may be bonding, nonbonding, or antibonding. Examples of this are seen in the spectrum of nitrogen shown in Fig. 3. Here the Π_u band starting at 16.7 eV results from the removal of the strongly bonding $\pi_u 2p$ electron. When an electron is removed from a bonding orbital, the equilibrium spacing of the nuclei of the ion is much greater than that of the molecule. Since the nuclei do not move much when the electron is removed, the ion is left in a state of high vibrational excitation. This is seen in the large number of observable vibrational levels in the Π_u band.

The probability that the transition occurs to any particular vibrational

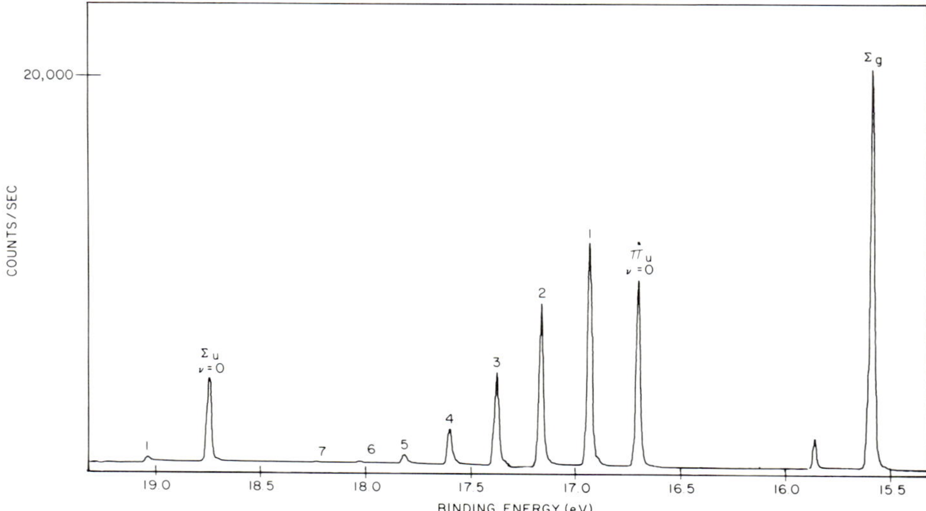

FIG. 3. Photoelectron spectrum of nitrogen.

state is determined by the Franck–Condon factors. These express the amount of overlap between the initial molecular state and the vibrationally excited ionic state. Transitions to the lowest vibrational level of the ion are called adiabatic transitions and transitions to the most probable vibrational level are called vertical transitions. The difference in energy between these two is a measure of the difference in spacing between the atoms of the ion and the molecule. The vibrational spacing of the Π_u band is 1810 cm^{-1}, significantly less than the 2331 cm^{-1} of the neutral molecule. This confirms that when this strongly bonding electron is removed, the atoms in the ion are less strongly bound to one another.

The Σ_g band at 15.6 eV shows only the slightest bonding character. Its vibrational spacing of 2100 cm^{-1} is only slightly less than that of the neutral molecule. The fact that only one vibrational level can be readily observed confirms the nonbonding nature of this orbital. The Σ_u band at 18.8 eV is slightly antibonding since its vibrational spacing is 2340 cm^{-1}, somewhat more than that of the neutral molecule.

Some other phenomena that are observed in diatomic molecules are the effects of autoionization and fragmentation. Autoionization occurs when an electron is removed from a lower lying orbital to a higher bound state of finite lifetime. This state then decays with the emission of an electron from an orbital lying higher in energy than the first orbital. These effects have been seen in nitrogen using neon radiation (Collin and Natalis, 1969) and in the angular dependence of emitted electrons (Carlson, 1971). Autoionization effects generally occur over a narrow photon energy range and have not been found to be very strong at 584 Å.

Fragmentation of the excited ion has been observed in the halogen acids (Turner *et al.*, 1970) and is seen by the loss of structure above $\nu = 3$ in the second band of HBr.

4 More Complex Molecules

In triatomic molecules, more than one set of vibrational modes may be excited. This gives rise to more complex vibrational structure in the photoelectron spectrum. For example, Eland and Danby (1968) have given a detailed vibrational analysis of SO_2. Turner *et al.* (1970) give interpretations for a large number of triatomic molecules. In more complex molecules, deuteration can be a useful technique in interpreting vibrational structure where hydrogen atoms are involved.

As the molecules become more complex, it becomes more important to have good molecular orbital calculations for comparison with experiment. Brundle *et al.* (1969) show how some corrections can be made to molecular orbital theory results to give better agreement with observed photoelectron spectra.

Other techniques are also used to identify features in spectra. For example, benzene is one of the most thoroughly studied of the slightly more complex molecules. The techniques that have been used to identify the features in the spectrum include of course molecular orbital theory (Jonsson and Lindholm, 1967), the effect of substituents (Baker *et al.*, 1968), measurement of intensities in the various bands (Asbrink *et al.*, 1970), and the angular dependence of electron ejection from different orbitals (Carlson and Anderson, 1971). The spectrum is shown in Fig. 4, with the orbital identification as given in Asbrink *et al.* (1970). Of the 30 valence electrons, 24 can be seen in the spectrum. The most readily identifiable vibrational structures can be ascribed to the symmetrical ring breathing mode. The small splitting in the first band at 9.25 eV is attributed to Jahn–Teller forces. This effect results from the fact that removal of an electron from certain symmetric molecules, such as methane (Bergmark *et al.*, 1972), destroys the symmetry of the molecule. The molecule then collapses to one of several less symmetric configurations where the electronic degeneracy of the original molecule is removed.

An example of the effect of substitution in benzene is given by chlorobenzene, shown in Fig. 5. The most obvious effect is the splitting in the

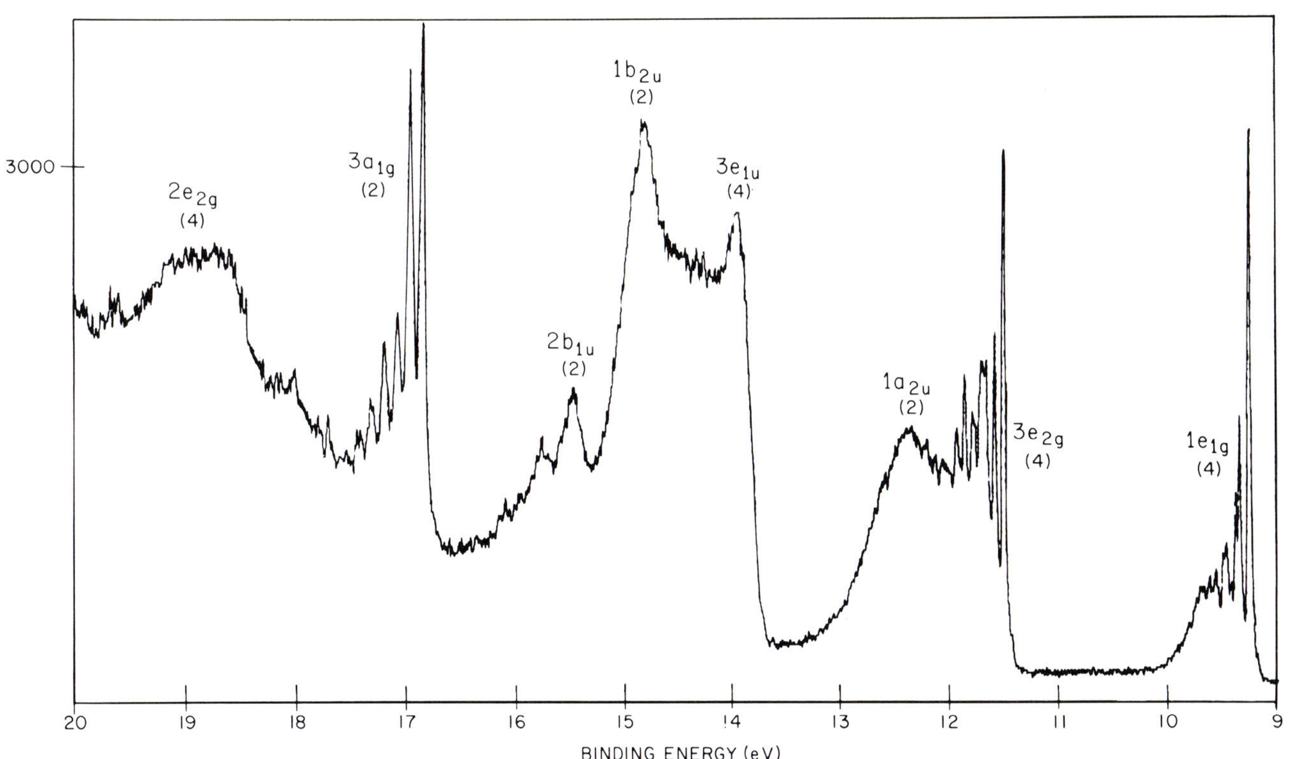

FIG. 4. Photoelectron spectrum of benzene. Numbers in parentheses give the number of electrons in the indicated orbital.

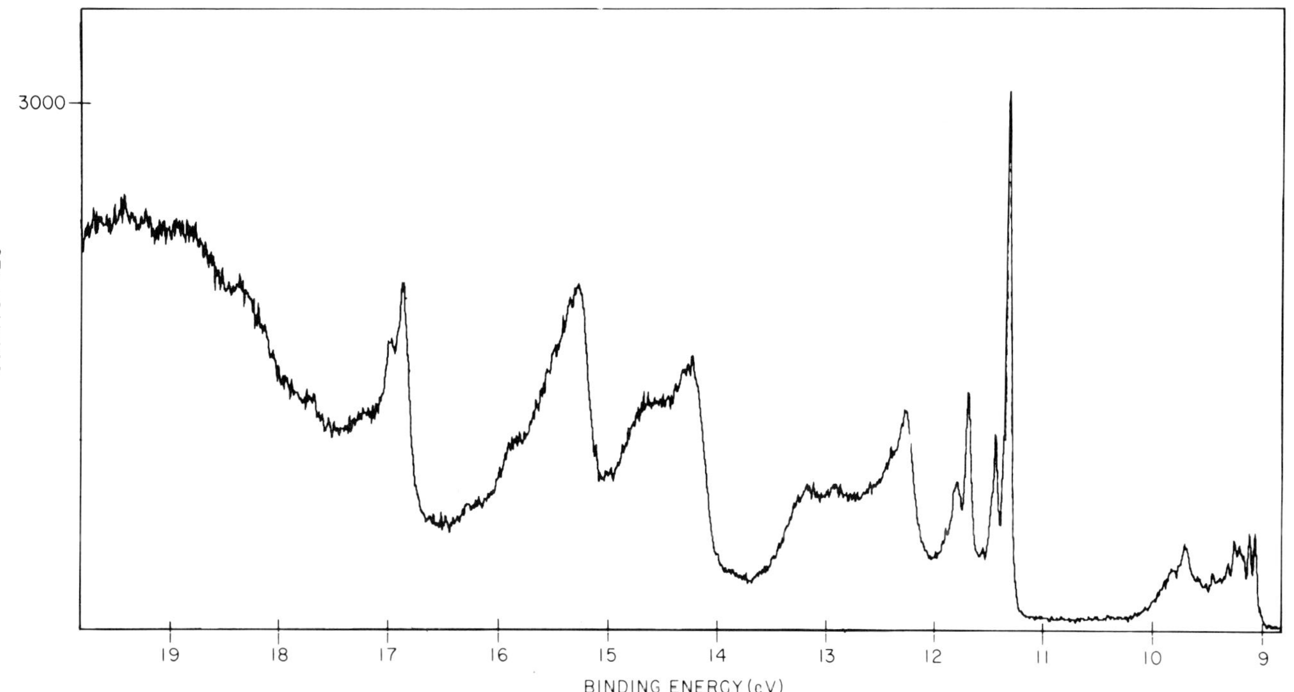

FIG. 5. Photoelectron spectrum of chlorobenzene.

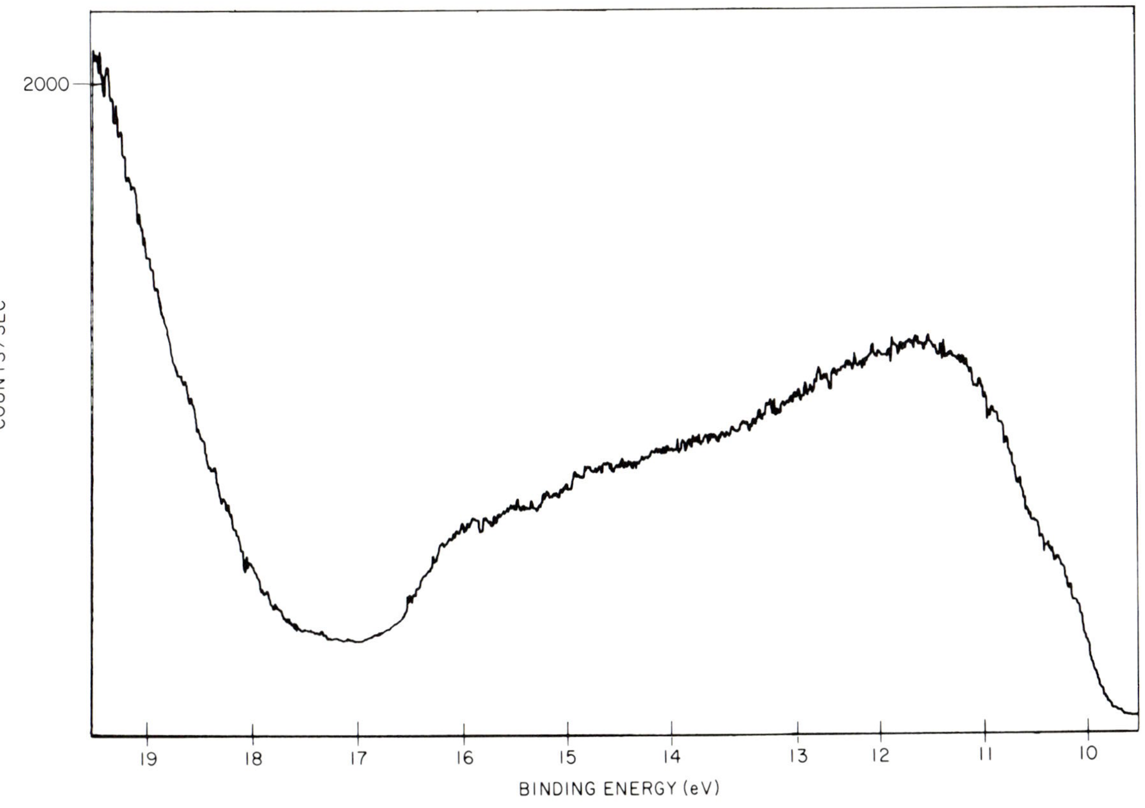

FIG. 6. Photoelectron spectrum of normal undecane.

$1e_{1g}$ band at 9.25 eV. This occurs because the chlorine reduces the molecular symmetry from D_{6h} to C_{2v} and this removes the degeneracy of the highest occupied π orbitals. The two peaks at 11.32 and 11.69 eV are associated with the chlorine lone pair electrons, the splitting being due to the differing interactions of the $3px$ and $3py$ orbitals with the benzene π electrons.

It seems that multiple bonding is generally necessary for there to be extensive vibrational structure and that saturated compounds often give spectra that show only broad diffuse bands. This is reasonable, since removal of an electron from a saturated band leaves the molecule loosely bound, with closely spaced vibrational levels. Also, more complex molecules will have more overlapping vibrational modes and more overlapping electronic states. Figure 6 shows an example of this kind of situation, where undecane ($C_{11}H_{24}$) shows no vibrational structure whatever.

Some of the other more complex molecules have spectra with more features in them. These may well be of value for fingerprint type identifications.

5 Analysis of Mixtures

The analysis of mixtures is certainly very possible with photoelectron spectroscopy. Because of varying photoelectric cross sections and different energy distributions of the spectra it is necessary for quantitative work to calibrate the spectrometer with mixtures of known composition. With samples of low volatility it is important that there be no selective distillation when the sample is vaporized. Figure 7 shows the spectrum of air. Because of its high cross section and the fact that all the photoelectrons appear in only two peaks, the argon (0.93%) in air is very detectable. The water vapor in laboratory air can be seen in Fig. 8. The main water (~1%) peak can be seen just to the left of the O_2 Π_g; $\nu = 2$ peak. Note the spin orbit splitting in the O_2 peaks. The N_2 Σ_g peak due to the second satellite of the helium lamp source at 23.74 eV can also be seen. It is almost impossible to see the CO_2 (0.03%) peak because of first-order satellite interference from N_2. The potential advantage of photoelectron spectroscopy in analyzing mixtures is in its ability to look at the electronic structure in cases where mass spectrometry is ambiguous, e.g., CO and N_2. It does not seem probable that its ability to see small concentrations in the presence of large major constituents will rival that of mass spectrometry.

There have been applications of mixture analysis in determining the electronic structure of transient species. For example, King *et al.* (1972) have measured CS present from a CS_2 electrical discharge and Daintith *et al.*

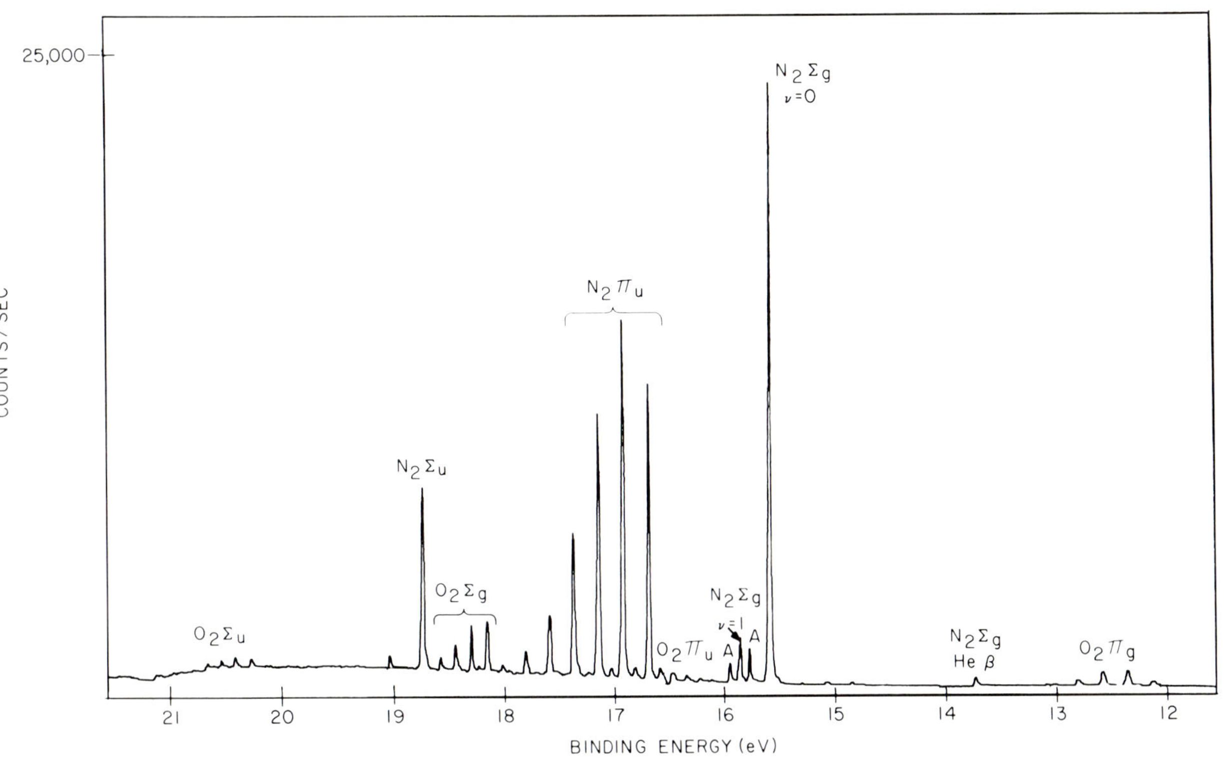

FIG. 7. Photoelectron spectrum of air.

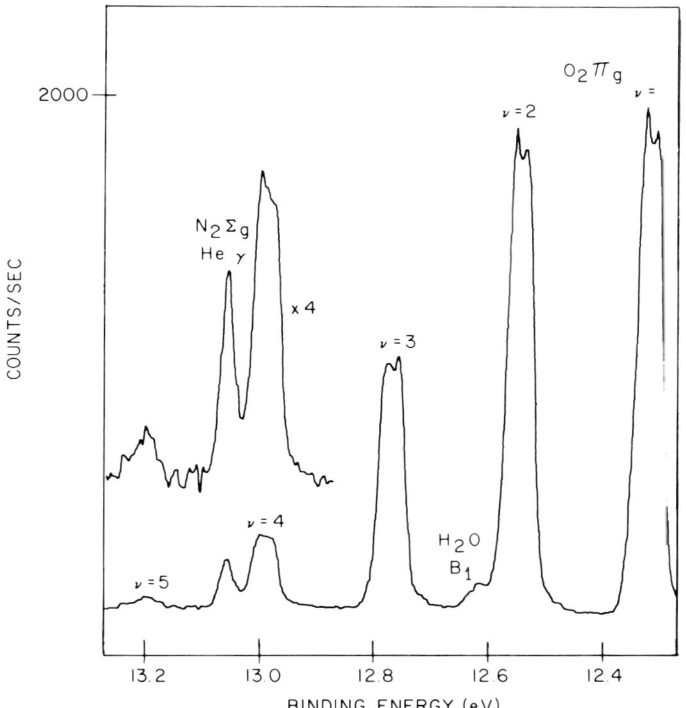

FIG. 8. Photoelectron spectrum of air, expanded to show the main water vapor peak.

(1972) have measured the decomposition products of the pyrolysis of formic acid.

6 Ultraviolet Photoelectron Spectroscopy of Solids

While there is little commercial instrumentation presently available, ultraviolet photoelectron spectroscopy of solid samples has been an important tool in the determination of band structure. The method is described by Derbenwick *et al.* (1974). Most of the work has been done using a LiF window which cuts off at 11.6 eV. Eastman and Cashion (1970) have shown how the windowless He lamps used in molecular photoelectron spectroscopy can be adapted to the ultrahigh vacuum conditions needed for work on solids and how the extra states due to species absorbed on a solid surface can be observed (Eastman *et al.*, 1971). The irradiated sample in this technique need only be a few millimeters square. Often the sample is evaporated since this is a simple means of producing an atomically clean surface.

7 Conclusion

Photoelectron spectroscopy has proved to be an extremely powerful technique for examining the electronic structure of matter. The high-resolution spectroscopy of ultraviolet excited vapors has been most important in giving an experimental check on molecular orbital theory. The analysis of gas mixtures promises to be important in studying the composition of reacting gases as well as the electronic structure of transient species involved in such reactions. Ultraviolet photoelectron spectroscopy of solids has given much good band structure data and in addition is quite sensitive to the presence of adsorbates on the surface.

Users of the technique have drawn upon the results of optical spectroscopy, electron impact ionization, and photon impact mass spectroscopy for comparison and confirmation of ionization potentials and vibrational structure. Because photoelectron spectroscopy furnishes data in a form that allows more direct comparison with theory, it is to be expected that it will supplant these other methods to some extent.

References

Al-Joboury, M. I., and Turner, D. W. (1963). *J. Chem. Soc.*, 5141.

Asbrink, L. (1970). *Chem. Phys. Lett.* **7,** 549.

Asbrink, L., Edquist, O., Lindholm, E., and Selin, L. E. (1970). *Chem. Phys. Lett.* **5,** 192.

Baker, A. D., May, D. P., and Turner, D. W. (1968). *J. Chem. Soc. B,* 22.

Bergmark, T., Rabalais, J. W., Werme, L. O., Karlsson, L., and Siegbahn, K. (1972). "Electron Spectroscopy" (D. A. Shirley, ed.), p. 413. North Holland Publ., Amsterdam.

Brundle, C. R., Turner, D. W., Robin, M. B., and Bosch, H. (1969). *Chem. Phys. Lett.* **3,** 292.

Carlson, T. A. (1971). *Chem. Phys. Lett.* **9,** 23.

Carlson, T. A., and Anderson, C. P. (1971). *Chem. Phys. Lett.* **10,** 561.

Collin, J. E., and Natalis, P. (1969). *Int. J. Mass Spectrosc. Ion Phys.* **2,** 231.

Daintith, J., Maier, J. P., Sweigart, D. A., and Turner, D. W. (1972). *In* "Electron Spectroscopy" (D. A. Shirley, ed.), p. 289. North Holland Publ., Amsterdam.

Derbenwick, G., Pierce, D., and Spicer, W. E. (1974). *Methods Exptl. Phys.* **11,** 67.

Eastman, D. E., and Cashion, J. K. (1970). *Phys. Rev. Lett.* **24,** 310.

Eastman, D. E., Cashion, J. K., and Switendick, A. G. (1971). *Phys. Rev. Lett.* **27,** 35.

Eland, J. H. D., and Danby, C. J. (1968). *Int. J. Mass Spectrosc. Ion Phys.* **1,** 111.

Jonsson, E., and Lindholm, E. (1967). *Chem. Phys. Lett.* **1,** 501.

King, G. H., Kroto, H. W., and Suffolk, R. J. (1972). *Chem. Phys. Lett.* **13,** 457.

Koopmans, T. (1934). *Physica* **1,** 104.
Lloyd, D. R. (1970). *J. Phys. E* **3,** 629.
Parker, J. H., and Warren, R. W. (1962). *Rev. Sci. Instrum.* **33,** 948.
Samson, J. A. R. (1969). *Rev. Sci. Instrum.* **40,** 1174.
Turner, D. W., Baker, C., Baker, A. D., and Brundle, C. R. (1970). "Molecular Photoelectron Spectroscopy." Wiley, New York.

CHAPTER 18

Visible and Ultraviolet Absorption Spectrometry

Richard S. Danchik

Aluminum Company of America
Alcoa Technical Center
Alcoa Center, Pennsylvania

Introduction

One of the most important functions of the analytical chemist is to determine qualitatively or quantitatively the chemical composition of various materials. These materials are virtually limitless in this respect. The selection of an analytical technique to provide the desired information about a given material is truly difficult, considering the vast selection of available analytical instrumental techniques, and variations in the physical state of the material and specimen size. The analyst must decide which instrumental technique will provide maximum results with a minimum expenditure of effort and within the constraints of specimen limitations.

This chapter discusses the practical application of visible and ultraviolet absorption spectrometry in modern analytical chemistry. The fundamentals of spectrophotometry will be reviewed, along with practical spectrophotometric methodology, and finally, specific analytical applications will be discussed and referenced.

The development of ultraviolet and visible absorption spectrometry as a discipline in modern analytical chemistry can be attributed primarily to the inherent utility of spectrophotometric methodology and the availability of reliable and inexpensive commercial spectrophotometers. No other methodology is as generally applicable in the determination of trace quantities of metals, nonmetals, and organic substances, although for specific applications, several other analytical methods are superior in respect to sensitivity, specificity, and speed.

1 Fundamentals of Spectrophotometry

Spectrophotometry is the science which deals with the measurement of the relative capacity of chemical systems to absorb incident radiant energy at specific wavelengths. Before considering methods of presenting and using spectrophotometric data, it is advisable to examine the terminology used in spectrophotometry. The recommendations on nomenclature and terminology outlined in *Analytical Chemistry* (1971) have been followed in this discussion and are listed in Table 1.

1.1 Origin of Ultraviolet and Visible Absorption Spectra

It is not within the scope of this chapter to discuss the principles of molecular spectroscopy involving group theory, quantum mechanics, etc. The analytical chemist is interested in the nature of the process for absorption spectra and the correlation of characteristic spectra with specific structures and substances. Therefore, only a brief description of the fundamentals of the absorption process will be given.

Radiant energy is the energy associated with electromagnetic waves of different wavelengths. It consists of minute units of energy called quanta, or photons. The relationship between the energy of a photon and the frequency appropriate for the description of its propagation is

$$E = h\nu \tag{1}$$

where E represents energy in ergs, ν represents frequency in cycles per

TABLE 1 RECOMMENDED SPECTROPHOTOMETRIC TERMINOLOGY[a]

Term	Symbol	Definition
Radiant power	P	The rate of transfer of radiant energy, i.e., radiant flux
Transmittance	T	The ratio of the radiant power transmitted by the sample P to the radiant power incident on the sample P_0, both being measured at the same wavelength and with the same slit width; $T = P/P_0$
Absorbance	A	The logarithm to the base 10 of the reciprocal of the transmittance; $A = \log(1/T) = -\log T$
Absorptivity	a	A constant characteristic of the absorptivity capacity of a specific absorber at a particular wavelength. The absorbance divided by the product of the concentration of the substance (in grams/liter, milligrams/milliliter, or micrograms/milliliter) and sample path length (in centimeters); $a = A/bc$
Absorptivity (molar)	ϵ	The absorptivity expressed in liter/mole cm when concentration is in moles/liter and sample path length is in centimeters. (If a is calculated using a concentration expressed in grams/liter, ϵ is equal to $a \times$ molecular weight of absorber)
Path length	b	Internal length of absorption cell, in centimeters
Concentration	c	Quantity of absorber per unit volume, moles/liter or gram/liter = milligram/milliliter, or milligrams/liter = microgram/milliliter = parts per million[b]
Resolution		The ratio of the average wavelength of two spectral lines which can just be identified as a doublet, to the differences in their wavelengths
Slit widths	SW	The slit width is the mechanical distance, in millimeters, between the sides of the narrow aperture which permits radiant energy to enter and leave the monochromator
	ESW	The width of the image of exit slit, along the wavelength scale, at which the radiant power is half of the maximum
	SSW (SRI)	The width of the image of the exit slit along the wavelength scale, i.e., the spectral region isolated
Bandwidth	OBW	Observed bandwidth is the width at one-half of the absorbance maximum as measured with a spectrophotometer
	NBW	Natural bandwidth is the width of one-half of the absorbance maximum due to a specific absorber. It is characteristic of absorber and not of instruments
Wavelength		The distance, measured along the line of propagation, between two points that are in phase on adjacent waves—units: angstroms, micrometers, nanometers, and microns

[a] From *Anal. Chem.* (1971). **43**, 2038. [b] For solvents with a density of 1.000 g/cm³.

second, and h is Planck's constant. Radiant energy also can be thought of as a continuous wave motion in which λ represents the interval between nodes in the wave pattern. The relationship between wavelength and frequency is

$$\nu = c/\lambda \tag{2}$$

where λ is the wavelength in centimeters and c is the velocity of the radiant energy (speed of light) in centimeters per second.

Visible light, a very small part of the electromagnetic spectrum, is generally considered to extend from 380 to 780 nm. The ultraviolet region of the electromagnetic spectrum is frequently subdivided into the far or vacuum ultraviolet region, approximately 10–200 nm, and the near ultraviolet region which extends from 200 to 380 nm. These classifications are arbitrary.

Table 2 presents an enlargement of the visible region, with the transmitted colors that correspond to various wavelengths. To state that a solution is colored means that of all wavelengths of white light incident upon the solution only selected wavelengths are absorbed, depending on the color of the solution; the remaining wavelengths are transmitted. A red solution, for example, appears red because it absorbs the shorter wavelength of the visible region and transmits the longer wavelengths; therefore, color is attributed to the selective absorption of incident radiant energy of certain wavelengths.

Absorption in the ultraviolet region of the spectrum has been related to

TABLE 2

RELATIONSHIP BETWEEN SELECTIVE LIGHT ABSORPTION AND COLOR

Observed (transmitted) (color of solution)	Wavelength region of maximum transmittance (nm)	Complementary (absorbed) hue (color of suitable glass filter)
Ultraviolet	<380	—
Violet	380–435	Yellowish green
Blue	435–480	Yellow
Greenish blue	480–490	Orange
Bluish green	490–500	Red
Green	500–560	Purple
Yellowish green	560–580	Violet
Yellow	580–595	Blue
Orange	595–650	Greenish blue
Red	650–780	Bluish green
Near infrared	>780	—

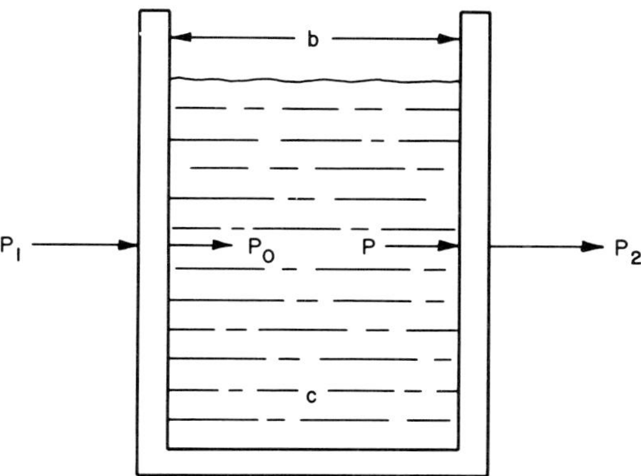

FIG. 1. Transmission of light through a solution: P_1, radiant power incident on first surface of medium; P_0, radiant power entering sample solution of concentration c and thickness b; P, radiant power leaving solution and incident on second surface; P_2, radiant power leaving last surface of medium.

the presence of unsaturation (multiple bonds or free radicals) and polarizability in ultraviolet absorbing materials. Saturated compounds are transparent in the ultraviolet region. Therefore, organic groups can be classified according to their effect on the ultraviolet absorption of the molecules to which they are attached. A chromophore is a group which, when introduced into a saturated hydrocarbon, produces a compound which has a selective absorption. The intensity of absorption of the simple chromophores (nitrite, carboxyl, etc.) varies widely from one group to the next. However, all members of a class of compounds containing a single chromophore will normally have absorption bands of approximately equal intensity and within a narrow wavelength range.

When an electromagnetic wave of a specific wavelength impinges upon a substance, the energy associated with that wave may be altered by reflection, refraction, absorption, and transmission processes. Reflection and refraction effects are generally negligible in the spectrophotometric analysis of solutions. The relationship between transmission and absorption processes is illustrated by Fig. 1. The rate at which energy is transported in a beam of energy is denoted by the symbol P_0 for the incident beam and by P for the quantity remaining unabsorbed after passage through a sample or container. The ratio of the radiant power transmitted by a sample to the radiant power incident on the sample is the transmittance T.

$$T = P/P_0 \tag{3}$$

The logarithm to the base 10 of the reciprocal of the transmittance is the absorbance.

$$A = \log_{10}(1/T) = \log_{10}(P_0/P) \tag{4}$$

1.2 Fundamental Laws of Absorption

There are two fundamental laws related to the absorption of monochromatic radiant energy by homogeneous, transparent media. The Bouguer–Lambert law expresses the relationship between the light absorptive capacity and the thickness of the absorber. Each layer of equal thickness will absorb an equal fraction of the beam of radiant energy which traverses it. Therefore, a beam of monochromatic radiant energy of radiant power P_0, upon passing through an absorber of thickness b, decreases in radiant power according to the expression (see Fig. 1)

$$dP = -k_1 P_0 \, db \tag{5}$$

Beer's law states that the fraction of the monochromatic radiant energy absorbed on passing through a solution is directly proportional to the concentration c of the absorber. This law can be expressed mathematically as

$$dP = k_2 P_0 \, dc \tag{6}$$

A combination of these two laws gives the exponential form

$$P = P_0 \exp(-k_3 bc) \quad \text{or} \quad \pm P_0 10^{-abc} \tag{7}$$

where $a = k_3/2.3$. In logarithmic form, this expression may be written

$$\log P_0/P = abc = A \quad \text{(absorbance)} \tag{8}$$

Absorbance is the product of the absorptivity a, the optical path length b, and the analytical concentration c.

This expression can be applied to analytical determinations by measuring A or P/P_0, the transmittance. When b is known, a can be calculated for a solution of known concentration at a specific wavelength. Knowing A, a, and b for an unknown solution, concentration c can be calculated provided that Beer's law is applicable.

1.3 Instrumentation

Ultraviolet-visible spectrophotometers consist of (1) an intense source of radiant energy in the 200- to 700-nm region, (2) a monochromator to isolate the wavelength region to be used in irradiating the solution, (3) an absorption cell assembly which provides for alternate examination of the reference solution and sample solution, (4) a photometer, comprised of a photoelectric detector which converts radiant energy to electrical energy, and a meter to indicate the resulting electric current. The distinguishing feature of a spectrophotometer is the use of a monochromator to select specific monochromatic radiant energy. From the practical operational viewpoint of the analytical chemist, the basic components of a spectrophotometer are reviewed.

1.3.1 *Sources*

The first essential component of a spectrophotometer is a source of continuous radiant energy at constant and sufficient intensity for the region of the spectrum in which the characteristic absorption bands on the sample are found. A tungsten filament lamp provides sufficient radiant power throughout the visible region and is widely used for the visible and near infrared region (350–3000 nm). Hydrogen or deuterium discharge lamps with quartz windows provide a continuous spectrum of radiant energy in the 185- to 370-nm region. Measurements below 195 nm require purging of the optical system with dry nitrogen to eliminate absorption due to oxygen and water vapor. Most commercial spectrophotometers are designed so that the ultraviolet lamp and the tungsten lamp, for visible spectrometry, can be rapidly interchanged.

1.3.2 *Monochromators*

The monochromator, the second essential component of a spectrophotometer, permits the selection of radiant energy of the desired wavelength. The monochromator consists of entrance and exit slits and a dispersive device, either a prism or grating, so arranged that radiant energy of a relatively narrow spectral bandwidth is obtained.

The capability of either a prism or a grating to separate two adjacent wavelengths is termed resolving power R. The resolving power of a prism is dependent on its effective thickness t and the slope of the dispersion plot of its optical material:

$$R = t(\partial n/\partial\lambda) \approx \lambda/\Delta\lambda \tag{9}$$

For a grating, the resolving power depends on the number of rulings N and the order of the spectrum m:

$$R = mN \tag{10}$$

Most monochromators used in commercial spectrophotometers have equal-width entrance and exit slits so that the energy distribution of the transmitted beam is triangular, as illustrated in Fig. 2. The slits control the spectral slit width (SSW) of the emerging beam, thereby causing the nominal wavelength to be slightly contaminated with radiant energy of

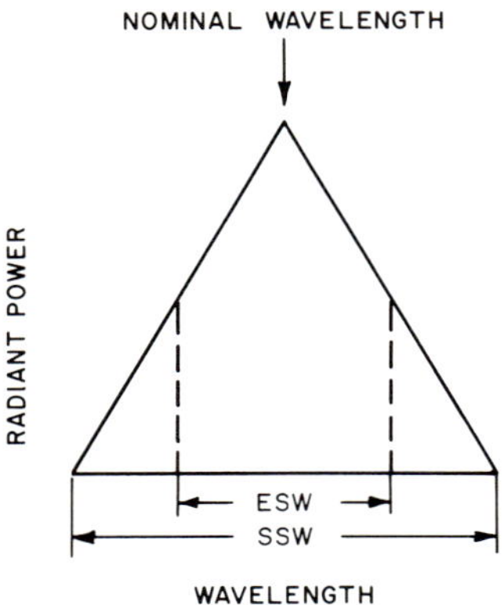

FIG. 2. Distribution of radiant energy transmitted by a monochromator having entrance and exit slits of equal widths.

slightly higher and lower wavelengths. The "effective spectral slit width" (ESW) is often used to designate that interval of the spectrum in which the radiant power of each wavelength is over half that of the maximum radiant power of the nominal wavelength.

1.3.3 *Absorption Cells*

The third essential component of a spectrophotometer is the assembly in which absorption cells are held and positioned in a reproducible manner. Absorption cells are constructed of glass for measurement in the visible region, while quartz or fused silica cells are used for measurement in the ultraviolet region. The sample and reference absorption cells should be matched with respect to optical path length and transmittance at specific wavelengths. It is especially important to purchase matched cells, since the transmittance characteristics in the ultraviolet region may be different, even though two cells might have identical lengths. Absorption cells in 1 cm internal optical path lengths are used extensively. Cells are also available with path lengths of 0.1, 0.5, 2, 5, and 10 cm.

1.3.4 *Photometers*

The most essential component of the photometer is the detector of radiant energy which produces a signal proportional to the radiant power of the beam. There are three types of photosensitive detectors commonly used: (1) barrier (photovoltaic) layer cells, (2) photoemissive cells, and (3) electron multiplier phototubes.

The photovoltaic cell consists of a conductor in close contact with a semiconductor. Electrons are transferred at the interface of these two layers when the semiconductor is irradiated. The electric current produced is proportional to the radiant power of the incident beam and to the area of the photosensitive surface being irradiated. Because of the low impedance of the cell, the output current cannot be amplified unless a cathode follower is used. Therefore, this type of detector is mainly used in filter photometers, where fairly high levels of illumination exist and where there is no need to amplify the signal. The spectral response of this detector is somewhat similar to the eye and is suitable for photometric measurements in the visible region.

The photoemissive cell consists of a photosensitive cathode containing an alkali metal oxide, and an anode mounted in a glass envelope, which is evacuated. The spectral response of this cell depends on the material used for the photocathode surface. Therefore, by proper selection of a photocathode the spectral response can correspond to the specific ranges of the visible and ultraviolet region. Radiant energy striking the photocathode causes photoelectrons to be emitted. The number of electrons is directly proportional to the radiant power of the incident beam, and the maximum velocity of the electrons is directly proportional to the frequency of the incident radiant energy. The high impedance characteristic of this phototube facilitates electronic amplification so that beams of very low radiant power can be measured.

The electron multiplier phototube (photomultiplier) has a series of photosensitive surfaces each charged at a successively higher potential, and the photoelectrons emitted by the first photocathode surface are accelerated from one dynode to the next, with the current being increased in each step by the secondary emission of electrons. The extreme sensitivity of this detector to low radiant power and its fast response time make it particularly suitable for high-resolution, recording spectrophotometers.

1.4 Practical Considerations in Selection of Instruments

The variety of commercially available spectrophotometers makes difficult the selection of an instrument to fulfill the function and service demands of a given laboratory. For routine determinations of a highly repetitive nature, an inexpensive filter photometer may be suitable. However, if new and untried spectrophotometric procedures are to be investigated, a high-resolution spectrophotometer is desirable. The cost of a spectrophotometer is closely associated with its resolution, photometric accuracy, and wavelength range. High-resolution instruments usually contain expensive optical components and complex electronic components necessary in a sensitive detector system. Numerous accessories are available for certain spectrophotometers to expand the basic capabilities of the instrument. Automatic sampling equipment, for instance, provides for measurements of numerous samples. Thermostated cells, provisions for plotting or direct digital readout in either absorbance or transmittance, and scale expansion over a narrower optical region are possible options. Provisions for flame emission, fluorescence, internal reflection, rapid scanning, dual wavelength, and photometric titration measurements are available with many instruments. The initial cost, availability of competent operators, adequate laboratory space, and proper maintenance procedures all should be weighed before the instrument is selected.

The mechanical, optical, and electrical systems constitute the principal design features to be considered in comparing commercial instruments used for spectrophotometric measurements. The option of using a filter or a monochromator to isolate and transmit a narrow band of radiant energy from the incident light determines whether the instrument is classified as a filter photometer or a spectrophotometer. Both instruments may employ either a single-beam or a double-beam optical system. With the single-beam instruments, a beam of radiant energy is first passed through a reference medium and the readout is adjusted to a 100% transmittance or zero absorbance setting. The beam then is passed through the sample to obtain the photometric reading. The stability of the source is vital for reliable photometric measurements with a single-beam instrument.

Double-beam instruments employ a beam-splitting device to produce two optical beams; one passes through the reference cell, the other through the sample cell. Some instruments have a detector for each beam so that the ratio of the two photocurrents being produced can be measured. Other instruments use a single detector which alternately receives the signal from the reference and sample beams. A rotating sector is often employed as a

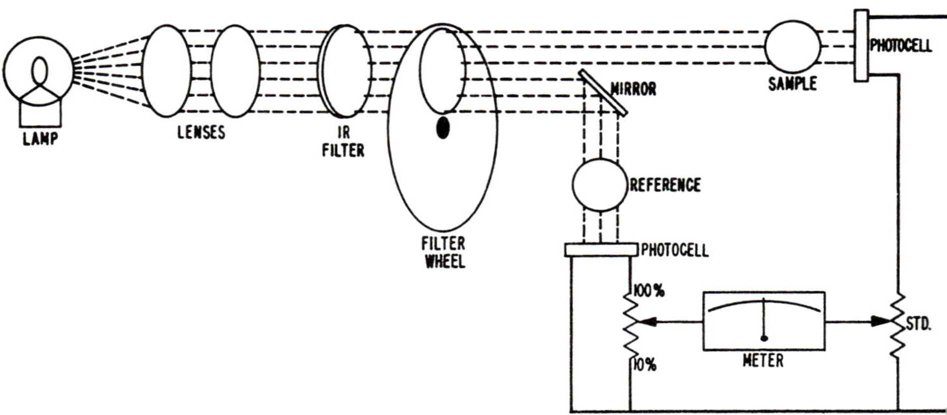

FIG. 3. Schematic diagram of Fisher Electrophotometer. [Courtesy of Fisher Scientific Company.]

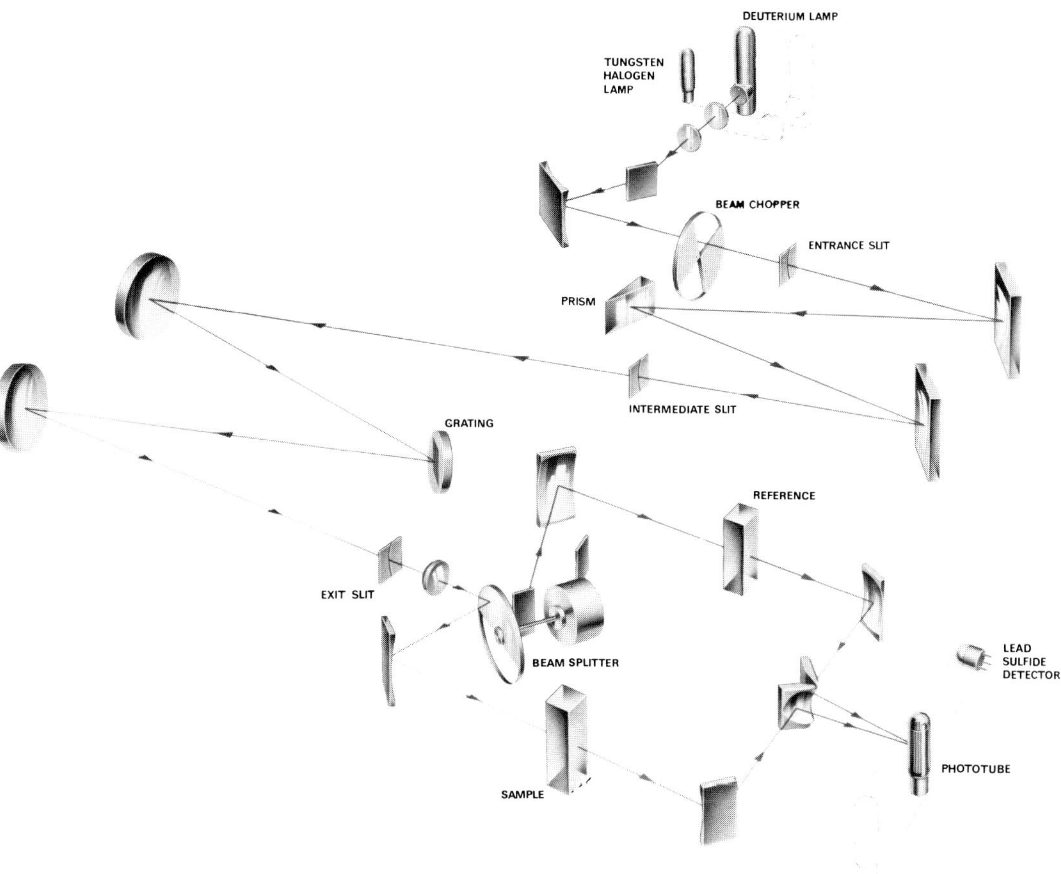

FIG. 4. Optical diagram of the Cary 17 Recording Spectrophotometer. [Courtesy of Cary Instruments.]

beam splitter in spectrophotometers which alternately transmits and reflects the incident monochromatic beam, but does so at a definite frequency. An ac amplifier tuned to this chopping frequency gives a high amplification with a minimum of noise. Schematic diagrams of typical absorptimetric instruments are shown in Figs. 3 and 4.

2 Spectrophotometric Methodology

2.1 Preparation of the Chemical System

The preparation of a suitable colored or absorption system for spectrophotometric measurements requires considerable care. First, the resultant system should be stable and capable of being prepared in a reproducible manner. Second, the absorbance of the system should be sensitive to small changes in concentration. For these conditions to be realized, it is necessary to control solution variables such as pH, concentration of reagents, and time and temperature required for development of maximum absorbance. In the selection of a photometric method and the development of a satisfactory absorptive system, the following criteria should be considered: sensitivity, reproducibility, stability, specificity, and effect of other substances likely to be present.

The preliminary treatment of a sample often requires dissolution by fusion or acid treatment. Although details of recommended chemical procedures cannot be covered here, it is essential that every effort be made to avoid contamination, since spectrophotometric methods are especially sensitive to traces of contaminants. Blank determinations should be made to compensate for any trace contaminants in reagents and other interferences inherent to the chemical system.

The composition of the matrix of the sample may make it necessary to isolate the desired constituent prior to measurement or to remove interfering substances. In analyzing inorganic systems, various methods of analytical separations are applicable such as ion-exchange, extractions, and electrodeposition methods. Distillation, precipitation, and various chromatographic methods also are used.

"Masking" is another technique to eliminate interferences by suppression of an undesired reaction employing a complexing agent which preferentially reacts with the interfering ion. Ringbom's (1963) book should be consulted for theoretical and practical aspects of complexation. In all cases, the preparatory treatment must be consistent for all samples and standards to ensure reproducible and accurate analytical results.

The following suggestions should be helpful in preparing a sample for spectrophotometric measurement:

(1) Know the selectivity and specificity of the reagents used for the preparation of a light absorptive system. Be aware of interferences that might be inherent in the sample itself.

(2) Determine the proper order of addition of reagents and allow time for completion of chemical reactions.

(3) Compensate for trace impurities in reagents by preparing a blank reference solution and by following identical procedures used in preparing sample solutions.

(4) If the sample solution exhibits an absorbance exceeding ~0.8, use a smaller sample size, dilute to a larger volume, or employ absorption cells of shorter path length in order to minimize the relative error in spectrophotometric measurements. The differential technique is an alternative approach to the measurement of high-absorbance systems.

(5) If the sample solution exhibits low-absorbance readings (less than 0.2), use a larger sample size, smaller final volume, longer optical path length absorption cells, or chemical concentration techniques.

(6) Absorption cells should be scrupulously cleaned, matched optically, and properly positioned in the optical beam of the instrument.

(7) For measurements in the ultraviolet region, select solvents having a transparency in the vicinity of the wavelength of measurement. Table 3 lists suitable solvents for ultraviolet measurement.

(8) Be sure that the solution is free of gas bubbles and turbidity before absorption measurements are made.

2.2 Measurement of the Absorptive System

After preparation of a suitable ultraviolet or visible light absorptive system, the next step in the general analytical procedure is to measure the absorptive capacity or absorbance at a specific wavelength. The selection of an appropriate wavelength for measurement should be based on a thorough knowledge of the chemical system being measured and the instrument being used. For most purposes, photometric measurements are made at the wavelength of maximum absorbance (absorption maximum) for the sample solution. However, there are specific instances under which measurements at the absorbance maximum are not recommended. Conformity to Beer's law is often observed at several different wavelengths, and it may be advantageous to measure at a wavelength where the system is not so sensitive, especially when high-absorbance systems are encountered. When slight variations in solution parameters cause differences in the absorbance maximum, it may be preferable to make measurements at a

TABLE 3

Selected List of Solvents for Ultraviolet Spectrometry

Solvent	Lower wavelength cutoff (nm, mμ)[a]	Solvent	Lower wavelength cutoff (nm)[a]
Acetic acid	270	Ethyl acetate	260
Acetone	330	Ethyl formate	265
Acetonitrile	212	Glycerol	230
Amyl acetate	260	*n*-Hexane	210
Benzene	280	*n*-Heptane	210
Butanol	220	Isooctane	210
n-Butyl acetate	260	Methanol	210
Carbon tetrachloride	260	Methylcyclohexane	210
Chloroform	240	Methyl formate	260
Cyclohexane	210	Isopropyl alcohol	210
1,2-Dichloroethane	235	Pyridine	300
Dichloromethane	233	Sulfuric acid (96%)	210
Diethyl ether	220	Tetrachloroethylene	295
N,*N*-Dimethylformamide	270	Toluene	285
p-Dioxane	220	Water	210
Ethanol	220	*m*-Xylene	290

[a] $b = 1$ cm.

shoulder on the spectrophotometric curve or at another absorbance band that is not so sensitive to solution variables. Finally, the avoidance of interferences due to another absorbing species at the sample absorbance maximum can be attained by selecting a wavelength where the interferer does not absorb, but the desired constituent has appreciable absorptivity.

For spectrophotometric measurements by the conventional method, the lower limit of the transmittance scale is set by adjusting the photometer readout to 0% T with the detector protected from all possible light sources. The upper limit of the transmittance scale is set by adjusting the photometer to read 100% T with only the solvent in the reference absorption cell. The 0% T reading is commonly referred to as the dark-current setting. After the dark-current and 100% T settings have been made, the sample solution is placed in the sample cell and the percent transmittance or absorbance is measured.

2.3 Presentation of Spectrophotometric Data

Spectrophotometric data are commonly presented in graphic form by (1) plotting either transmittance or absorbance versus wavelength in

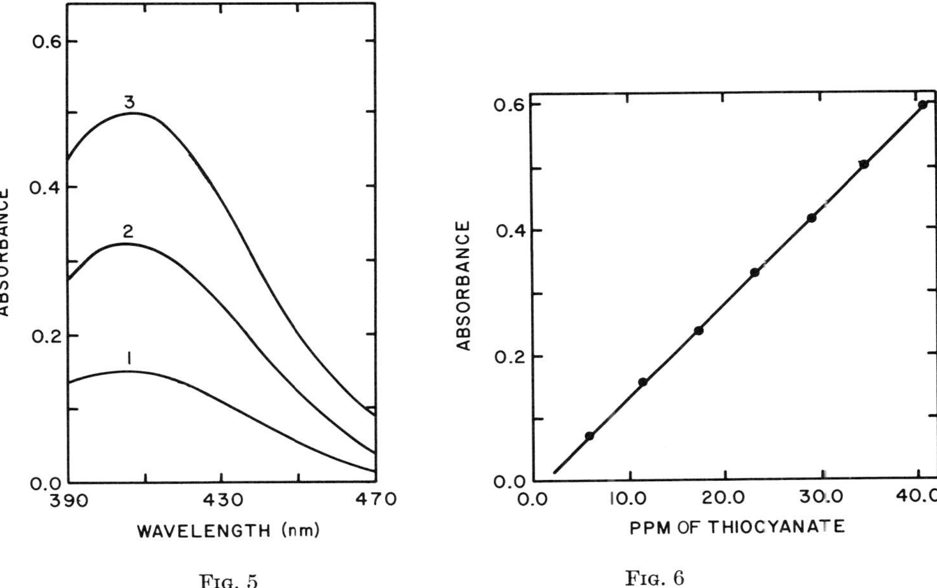

FIG. 5

FIG. 6

FIG. 5. Typical absorbance versus wavelength plot [absorption spectra for dithiocyanatodipyridine copper(II)—Curve 1: 11.6 ppm SCN; curve 2: 23.2 ppm SCN; curve 3: 34.8 ppm SCN].

FIG. 6. Plots testing for conformity to Beer's law [plot for dithiocyanatodipyridine copper(II)].

nanometers (millimicrons) to obtain the characteristic absorption spectrum of the absorber (Fig. 5), and (2) plotting absorbance or log $1/T$ values versus concentration in order to obtain a calibration graph (Fig. 6). A linear relationship for an absorbance versus concentration plot indicates conformity to Beer's law.

In practice, it is always advisable to specify the wavelength of measurement, the solution used in the reference cell, the thickness of the cell, the spectral bandwidth, the temperature of the solution, the instrument used for measurement, and any other significant variables.

2.4 Sources of Error

Most errors in spectrophotometric analysis can be attributed to three major factors: (1) the nature of the chemical system prepared for measurement, (2) the operational characteristics of the instrument being used, and (3) faulty analytical techniques.

Chemical effects which can cause errors in absorbance measurements include association, dissociation, shifts in equilibrium with changes in ionic strength and pH, solute–solvent interaction, and consumption of reagent by diverse ions. The accuracy of a spectrophotometric analysis in many respects depends on representative sampling (especially in trace analysis), sample aliquot size, the character and properties of the material being analyzed, and the accuracy in the preparation of suitable standard solutions. The absorptive system should be a homogeneous solution, free of turbidity. The stability of the solution must be sufficient to permit photometric measurements within a reasonable time period. The skill of the analytical chemist, his depth of understanding of all of the chemical processes in the course of an analysis, and his manipulative skill in the chemical preparation of an absorptive system all determine the accuracy of a spectrophotometric method.

Spectrophotometric measurements are subjected to three instrumental errors: (1) finite slit width effect, (2) the multiple reflection path effect, and (3) the stray radiant energy effect. Instrumental errors depend on the instrument being used and the manner in which it is being operated.

2.4.1 *Finite Slit Width Effect*

The minimum spectral slit width which can be achieved with a given spectrophotometer is related to the spectral emittance curve of the source,

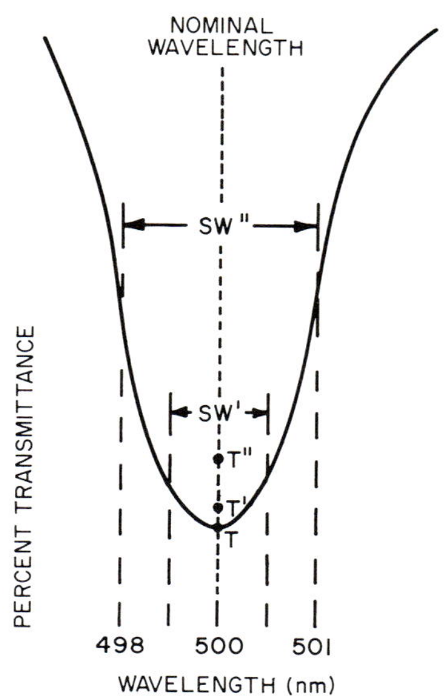

FIG. 7. Diagram illustrating finite slit width effect—T' and T'' show the deviation of the measured transmittance values from the true T value as a function of the slit width being used.

the spectral response curve of the phototube, and the resolving power of the monochromator. The finite slit width effect refers to this minimum spectral slit width and the extent of this error depends on the spectral bandwidth and the shape of the absorption band. Therefore, only when the absorbance maximum of the spectrophotometric curve is very narrow at the wavelength of measurement, is an error likely to result. Figure 7 illustrates the nature of this potential error.

2.4.2 *Multiple Reflection Path Effect*

This error occurs in a solution exhibiting low absorbance. It is caused by the optical beam being reflected repeatedly by the front and back surfaces of the absorption cells and to a lesser extent by the exit slit lens and faces of the slit jaws. Part of the incident radiant energy traverses the absorption cell more than once, resulting in too high an absorbance value. This error can be minimized by using longer cell lengths and antireflection coatings (Goldringer *et al.*, 1953).

2.4.3 *Stray Radiant Energy Effect*

The reflection and scattering of light at surfaces of the optical components of the monochromator can introduce stray radiant energy which is superimposed on the monochromatic beam emerging from the instrument. This error occurs in measuring solutions of high absorbance ($A > 1.5$) and a negative deviation from Beer's law is observed (Slavin, 1963; Miranda and Conte, 1971).

Faulty analytical techniques might include unclean absorption cells, measurement outside the optimum concentration range, measurement on the steep segment of a spectrophotometric curve, dilution errors, misalignment of the source so that wide slit widths must be used, and volumetric changes due to temperature.

In selecting a spectrophotometric method for determining a specific constituent, the approximate concentration range of the constituent should be considered. The optimum concentration range for most spectrophotometric methods corresponds to solutions having an absorbance in the 0.2–0.7 region. To determine the optimum concentration range for a specific method and a specific instrument, prepare a plot of percent transmittance versus the logarithm of the concentration (Ringbom, 1939). This Ringbom plot (see Fig. 8) has a virtually linear segment of the curve corresponding to the optimum concentration range. With the use of a precision spectrophotometer and exceptional precautions, photometric errors corresponding to 0.1 to 0.2% may be achieved; however, a more reasonable estimate for most spectrophotometric measurements would be 0.5%. If errors in preparing the sample are negligible, a relative error of less than 2% should be obtainable

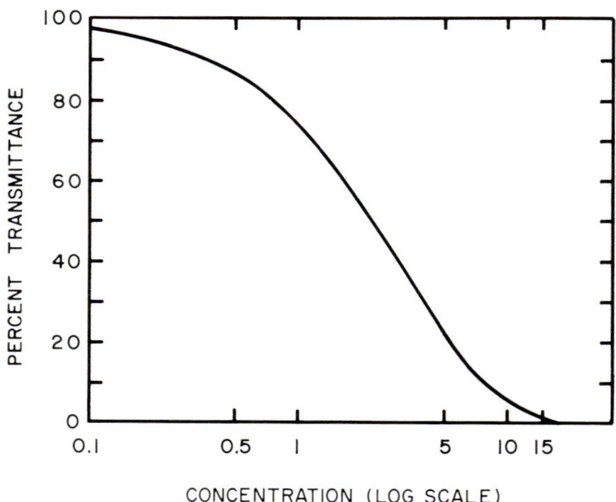

FIG. 8. Ringbom plot—optimum concentration range. The concentration is expressed in parts per million.

by confining measurements to the 20–60% transmittance (0.2–0.7 absorbance) region. Sandell (1959) proposed a convenient method of expressing sensitivity in terms of the gamma (10^{-6} g) of an element per milliliter of solution which gives an absorbance change of $0.001A$ (approximately 0.2% in transmittance).

2.5 Special Spectrophotometric Techniques

2.5.1 *Differential Spectrophotometry*

In the analysis of very concentrated or very dilute solutions, photometric error can be minimized by using a differential technique, commonly called the "transmittance ratio method." Figure 9 illustrates the scale expansion by this differential technique. In this method, the 0% T setting is made with the phototube in total darkness, and the 100% setting is adjusted using a standard solution slightly less concentrated than the most dilute sample solution. Differential techniques have been used extensively in metallurgical analysis to obtain relative standard deviations of less than 0.1–0.2% (Bastian *et al.*, 1949, 1950). Figure 9 also illustrates the "trace analysis" and "maximum precision" differential techniques.

2.5.2 *Multicomponent Analysis*

It is possible to quantitatively determine two constituents in a sample by making photometric measurements at two wavelengths. The proper conditions for this technique require that each constituent have a character-

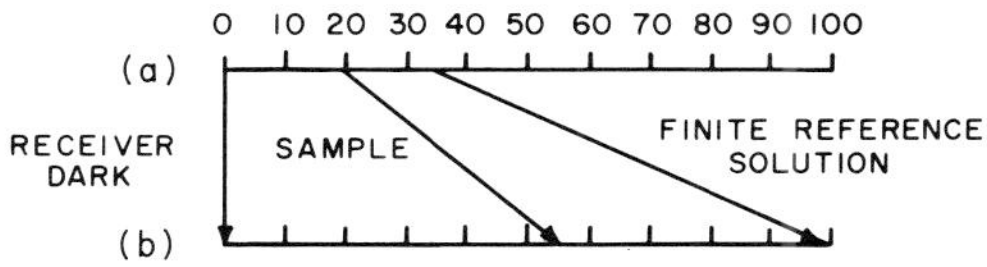

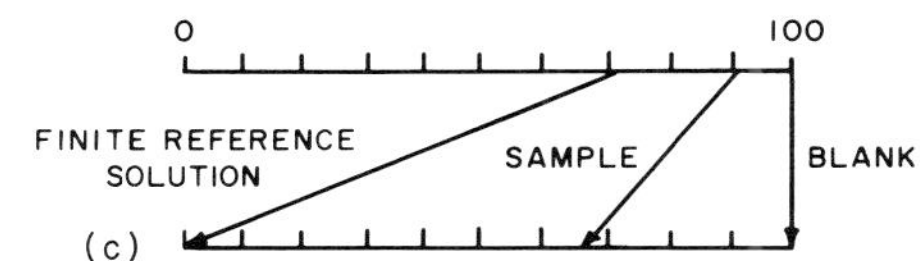

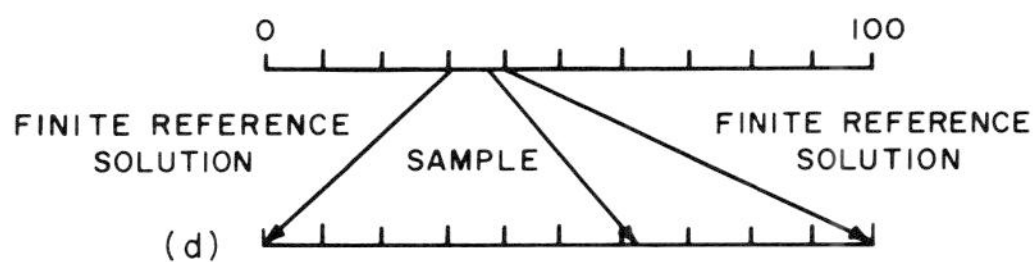

FIG. 9. Differential spectrophotometric techniques: (a) conventional method; (b) transmittance ratio method; (c) trace analysis method; (d) maximum precision method.

istic absorption maximum at a wavelength where the other constituents have low absorbances, and that Beer's law is obeyed for each constituent at the two wavelengths. The following procedure summarizes the spectrophotometric analysis of a binary system as proposed by Boltz (1952).

(1) Determine the spectrophotometric curves for the standard solutions of each component. Use the same reference solution.

(2) Determine the two wavelengths at which there is a maximum in the difference of absorbances.

(3) Test each component for conformity to Beer's law at the two selected wavelengths.

(4) Calculate the absorptivity a for each constituent at both wavelengths.

(5) Prepare mixtures of the components and measure absorbances at the two wavelengths. Plot the observed absorbances at each wavelength versus the calculated absorbances. If additive, a linear plot is obtained.

(6) Solve the following simultaneous equations for c_1 and c_2.

$$\begin{aligned} A_1 &= a_1{}^{\mathrm{I}}bc_1 + a_1{}^{\mathrm{II}}bc_2 \\ A_2 &= a_2{}^{\mathrm{I}}bc_1 + a_2{}^{\mathrm{II}}bc_2 \end{aligned} \tag{11}$$

2.5.3 *Indirect Techniques*

Indirect spectrophotometric methods are based on two specific techniques: (1) measuring the absorbance of a specimen equivalent to the amount of the desired constituent, and (2) measuring the decrease in absorbance of a system due to a chemical reaction with the desired constituent. The latter technique is often referred to as a "bleaching effect." For example, thallium may be determined indirectly based on the precipitation of thallium (I) molybdophosphate, dissolution in a borate buffer solution, then measurement of the ultraviolet absorbance of the equivalent molybdate (Hargis and Boltz, 1965). The bleaching effect of fluoride on the zirconium-eriochrome cyanide R complex serves as the basis of an indirect spectrophotometric method for fluoride.

2.5.4 *Photometric Titration*

A photometric titration may be defined as a change in absorbance of a solution used to follow the change in concentration of a light-absorbing constituent during a titration. A plot of absorbance versus the volume of titrant will consist of two straight lines intersecting at the titration end point—similar to amperometric and conductometric titrations. Photometric titrations have several distinct advantages over direct spectrophotometric determinations. The presence of other absorbing species at the analytical wavelength does not necessarily cause an interference, since only the change in absorbance is significant. The only condition required is that the absorbance of the light-absorbing species should be considerably greater than that of impurities in the solution. A second advantage is that only a single absorber need be present from among the titrant, reactant, or the reaction products. This extends spectrophotometric methods to a large number of nonabsorbing constituents. A typical example of a photometric titration would be in the determination of magnesium in aluminum alloys using EDTA as the titrant and chrome azurol S as the indicator (Kanie, 1957). More detailed information is given for photometric titrations by Headridge (1961).

3 Analytical Applications

3.1 Special Applications of Spectrophotometry

Among the various applications of spectrophotometry is the determination of empirical formulas of complexes formed by metal ions and ligands. There are three spectrophotometric methods used for the determination of the composition of complexes: (1) mole-ratio method, (2) slope-ratio method, and (3) continuous variations method.

3.1.1 *Mole-Ratio Method*

Consider the complex M_xL_y, which has a characteristic absorption maximum; the ligand-to-metal ratio can be determined spectrophotometrically by the mole-ratio method provided that the complex has a sufficiently large formation constant K_f. It is also necessary that the metal and complexing reagents do not exhibit an absorbance at the absorption maximum for the complex. Figure 10 illustrates the mole-ratio method of identifying complexes. In practical application, a series of solutions is prepared in which the metal ion concentration C_M is maintained constant and the concentration of the ligand C_L is varied. A plot of the ligand-to-metal ratio, C_L/C_M, versus absorbance shows an intersection of a gradually increasing line and a virtually horizontal line indicating attainment of maximum absorbance for the complex. The intersection of these two lines is indicative of the composition of the complex.

3.1.2 *Slope-Ratio Method*

In this method two series of solutions are prepared. The first series contains a constant concentration of ligand C_L which corresponds to a sufficient excess to minimize any appreciable dissociation of the complex. Concentrations of the metal ion C_M are varied so that the absorbance of the complex can be assumed to be proportional to the metal ion concentration C_M.

$$C_{M_xL_y} = C_M/x \tag{12}$$

A plot of absorbance versus C_M is linear with the slope equal to $1/x$.

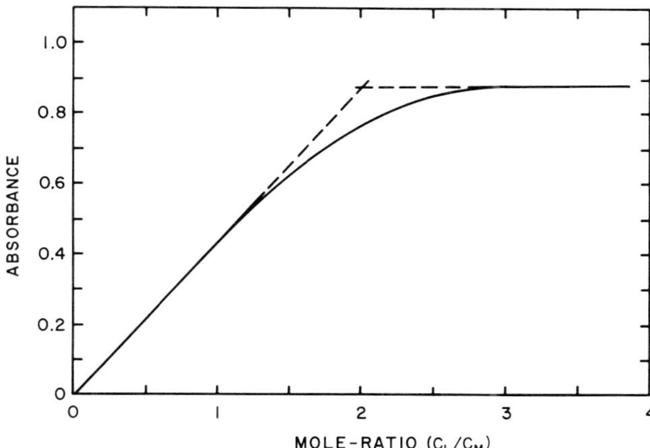

FIG. 10. Mole-ratio method of identifying complexes.

A second series of solutions is prepared containing a constant concentration of metal ion and various concentrations of the ligand. The following relationship is obtained:

$$C_{\mathrm{M}_x L_y} = C_{\mathrm{L}}/y \qquad (13)$$

when the metal ion is sufficiently large to insure that the absorbance is proportional to the ligand concentration C_{L}. The slope is obtained from a plot of the absorbance versus C_{L} and equal to $1/y$. The ratio of the slopes for the two plots is equal to the ligand-to-metal ratio y/x. Figure 11 illustrates the slope-ratio method of identifying complexes.

3.1.3 *Continuous Variations Method*

Since its introduction by Job (1928), the method of continuous variations has been used most often to determine the formulas and formation constants of complex ions in solution. The method is based on plotting measured absorbances, corrected for absorbances of reactants assuming no complexation, versus the mole fraction of either the ligand or metal. The method requires the preparation of a series of solutions in which the molar concentrations of metal and ligand are varied, but their total molar concentration, $C_{\mathrm{M}} + C_{\mathrm{L}}$, remains constant. The absorbance versus mole fraction plot gives a characteristic triangular plot as shown in Fig. 12. The mole fraction of the maximum of this plot, the apex of the triangle, indicates the composition of the complex.

A "normalized absorbance" concept has been recently utilized in the continuous variations method by Likussar and Boltz (1971a). General equations for the calculation of conditional extraction and formation

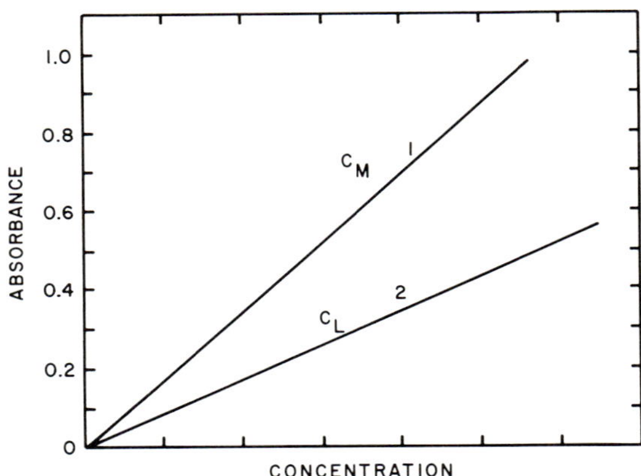

FIG. 11. Slope-ratio method of identifying complexes.

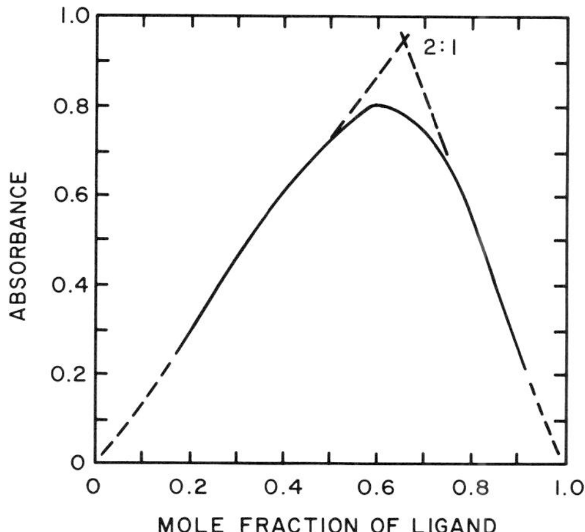

FIG. 12. Continuous variation method of identifying complexes.

constants have been developed and the limiting parameters for the calculation of these constants have been delineated. Modified spectrophotometric methods have been used to determine the extraction constants for 1-pyrrolidinecarbodithioate complexes of Cu(II), Co(II), Cd(II), Zn(II), Bi(III), and Ga(III). The formation constant of 1-pyrrolidinecarbodithioic acid and the distribution ratio at various pH values have been evaluated (Likussar and Boltz, 1971b).

3.1.4 *Determination of pK Values*

In accordance with the theory of pH indicators, it is known that a state of equilibrium exists between the two molecular forms of the indicator, that each molecular form has a characteristic color, and that the relative amount of each molecular form depends upon the pH of the solution. With this knowledge, the pK value of an acid–base indicator can be determined spectrophotometrically. A series of solutions of known pH containing a constant total concentration of indicator C_{IND} is prepared. These solutions should be of known and constant ionic strength. The absorbance is measured at the wavelength of the absorption maximum and a plot of absorbance versus pH is obtained (see Fig. 13). The inflection point of this plot, $(A_{\max} - A_{\min})/2$, corresponds to the p$K \pm 2$ range of the indicator.

$$\mathrm{pH} = \mathrm{p}K + \log\,(\text{proton accepter/proton donor}) \tag{14}$$

The wavelength at which all solutions have an identical absorbance is designated as the "isosbestic point." At this point, the absorbance for both

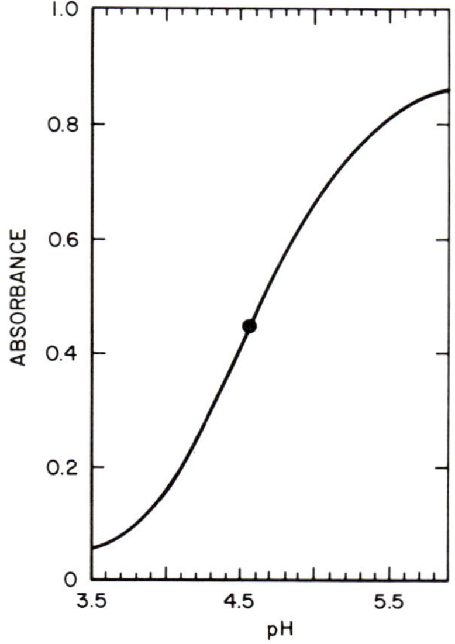

FIG. 13. Spectrophotometric determination of pH using bromcresol green indicator.

molecular forms of the indicator is the same and only dependent on the concentration of the indicator.

3.2 Analysis of Metals

The spectrophotometric method for the determination of metals is one of the most extensively employed analytical techniques because it is sensitive, specific, rapid, and the instrumentation required is relatively inexpensive. Organic reagents which form colored metal chelates are frequently utilized for metal analysis. Often the sensitivity of these spectrophotometric methods can be increased by extracting the metal chelate into a relatively small volume of an organic phase. Sensitivities, selectivities, and sample size for spectrophotometric methods vary according to the individual procedure and cannot be generally characterized. Most spectrophotometric methods are destructive in that the sample is changed by chemical means, e.g., dissolution, complex formation, etc. However, many methods can use the sample directly if it is a liquid.

It would be impossible to summarize the numerous spectrophotometric methods available today. In any analytical chemistry journal, new and improved procedures for spectrophotometric analysis are continually reported. Treatises by Sandell (1959), Snell and Snell (1959, 1967, 1970); and Charlot (1964) contain numerous colorimetric methods for metals and procedures for specific applications to particular materials including

TABLE 4

SELECTED SPECTROPHOTOMETRIC METHODS FOR THE DETERMINATION OF METALS

Metal	Reagent or method	References
Al	Catechol violet	Mustafin *et al.* (1967)
	Alkalone	Danchik and Oliver (1970)
Au	Ammonium-1-pyrrolidine dithiocarbamate	Berger *et al.* (1971)
Be	Chrome Azurol S	Sommer and Kuban (1968)
B	Curcumin	Tolk *et al.* (1969)
Ca	Glyoxal bis(2-hydroxyanil)	Lapid and Pickholtz (1969)
Cu	Polyethylenimine	Ziegler and Ziegeler (1967)
Fe	1, 10-Phenanthroline	Stephens and Suddeth (1967)
Ga	1-Pyrrolidine carbodithioate	Likussar *et al.* (1970a)
Pb	Hydroxyhydroquinone	Wawrzyczek and Polak (1967)
Mg	Chromotrope 2R	Shibata *et al.* (1969)
Mn	Periodate	Jacobson (1968)
Ni	Dimethylglyoxime	British Standards Institute (1969)
Ti	Tiron	Clark (1970)
Re	Pyrrolidine dithiocarbamate	Likussar *et al.* (1970b)
W	Peroxytungstic acid	Parker and Boltz (1968)

minerals, ores, and water. A comprehensive summary of ultraviolet-visible methods for metals, nonmetals, and organic substances is given by Boltz and Schenk (1963). New and improved methods for both ultraviolet (Crummett and Hummel, 1966, 1968, 1970), and visible (Boltz and Mellon 1964, 1966, 1968, 1970) spectrophotometry are documented periodically in biennial reviews. Also, specific applications for various analytical methods including spectrophotometry are contained in biennial reviews, supplying up-to-date information in such fields as air and water pollution, clinical chemistry, ferrous and nonferrous metallurgy, and other areas. (Annual Reviews, 1969, 1971). *Scott's Standard Methods of Chemical Analysis* (Furman, 1962), and ASTM standard procedures (1970) also are useful for the determination of metals, nonmetals, and organic substances in a variety of materials. Representative spectrophotometric methods which have been used in the analysis of metallurgical materials are cited in Table 4.

3.3 Analysis of Nonmetals

Spectrophotometry in the determination of nonmetals involves a variety of chemical reactions. Heteropoly complexes, dyes, ion association complexes, ternary complexes, and salts are the more common species giving measurable absorptive systems incorporating anions and nonmetals.

Nonmetallic elements such as phosphorous and silicon form yellow heteropoly complexes which can be measured photometrically, or these complexes can be reduced to produce heteropoly blues which have unusually high absorptivity values. Thiocyanate can be determined by the formation and selective solvent extraction of the dithiocyanatodipyridine copper(II) complex (Danchik and Boltz, 1968). Spectrophotometric methods for most of the nonmetals and the application of these methods to the analysis of numerous metallurgical samples have been discussed by Boltz (1950). Table 5 lists selected spectrophotometric methods for nonmetals.

3.4 Analysis of Organic Compounds

Ultraviolet spectrophotometry has been applied extensively to the determination and identification of organic substances such as aromatic hydrocarbons, vitamins, steroids, heterocyclics, and conjugated aliphatics. Ultraviolet absorption spectra are used often for the identification of degradation products and to test for purity in biochemical and pharmaceutical research. In qualitative analysis, the correlation of ultraviolet absorption bands with specific structures is made chiefly by analogy. Table 6 lists several common chromophoric groups and the appropriate wavelengths of maximum absorbance.

Collections of spectra and indexes of ultraviolet spectra are extremely useful in qualitative and quantitative analytical studies. The Sadtler Res. Lab., and Lang (1961–1965) collections of spectra are available. Several volumes entitled, *Organic Electronic Spectral Data* (Kamlet, 1960; Ungnade, 1960), the ASTM E-13 IBM cards (1957), and an index (Hershenson,

TABLE 5

Selected Spectrophotometric Methods for the Determination of Nonmetals

Nonmetal	Reagent or method	References
B	Curcumin	Grinstead and Snider (1967)
	1,1-Dianthrimide	Gupta and Boltz (1971)
ClO_2	Tris-1,10-phenanthroline iron(II)	Howell *et al.* (1970)
F	Zirconium–eriochrome cyanine R	Megregian (1954)
NO_2	Antipyrine	Weiss and Boltz (1971)
OH^-	Alkalone	Danchik and Oliver (1970)
P	Heteropoly blue	Lueck and Boltz (1956)
Si	Heteropoly blue	Pakalns and Flynn (1967)
SO_2	Direct method	Bhatty and Townshend (1971)
SCN	Copper(II), pyridine ($CHCl_3$)	Danchik and Boltz (1968)

TABLE 6

SELECTED LIST OF CHROMOPHORIC GROUPS

Chromophoric group	System	λ_{max} (nm)
Nitrile	$-C{\equiv}N$	<160
Acetylenic	$R-C{\equiv}C-R$	173
Ethylenic	$R-CH{=}CHR$	193
Azomethine	$>C{=}N-$	190
Carboxyl	$RCOOH_2$	204
Amido	$RCONH$	<208
Nitrate	$-ONO_2$	270
Nitro	$-NO_2$	271
Carbonyl (ketone)	$RR'C{=}O$	271
Carbonyl (aldehyde)	$RHC{=}O$	293
Nitroso	$-N{=}O$	300
		665
Azo	$-N{=}N-$	347
Nitrite	$-ON{=}O$	370

1956, 1961) are recommended for the identification of organic compounds. Several comprehensive books on interpretation of ultraviolet absorption spectra are available (Silverstein and Bassler, 1967; Scott, 1964).

Ultraviolet absorption is used extensively in the determination of molecular structures. Some of the more common applications include

(1) detection of a particular group;
(2) determination of the position in the molecule of a particular group;
(3) studies of steric effects,
(4) determination of cis–trans isomers,
(5) choice of correct structure among several possibilities, and
(6) quantitative determination of tautomerization equilibria.

References

American Society for Testing and Materials (1957). Committee E-13, *J. Opt. Soc. Amer.* **47,** 672.

Anal. Chem. (1971). **43,** 2038.

"Annual Book of ASTM Standards" (1970). American Society for Testing and Materials, Easton, Maryland.

Annual Reviews, Applications (1969). *Anal. Chem.* **41,** 1R-360R.

Annual Reviews, Applications (1971). *Anal. Chem.* **43,** 1R-388R.

Bastian, R., Weber, R., and Palilla, F. (1949). *Anal. Chem.* **21,** 972.

Bastian, R., Weber, R., and Palilla, F. (1950). *Anal. Chem.* **22,** 160.
Berger, A. S., Likussar, W., and Boltz, D. F. (1971). *Microchem. J.* **16,** 286.
Bhatty, M. K., and Townshend, A. (1971). *Anal. Chim. Acta* **55,** 263.
Boltz, D. F. (ed.) (1950). "Colorimetric Determination of Non-Metals." Wiley (Interscience), New York.
Boltz, D. F. (1952). "Selected Topics in Modern Instrumental Analysis," p. 105. Prentice-Hall, Englewood Cliffs, New Jersey.
Boltz, D. F., and Mellon, M. G. (1964). *Anal. Chem.* **36,** 256R.
Boltz, D. F., and Mellon, M. G. (1966). *Anal. Chem.* **38,** 317R.
Boltz, D. F., and Mellon, M. G. (1968). *Anal. Chem.* **40,** 255R.
Boltz, D. F., and Mellon, M. G. (1970). *Anal. Chem.* **42,** 152R.
Boltz, D. F., and Schenk, G. H. (1963). "Handbook of Analytical Chemistry" (L. Meites, ed.), p. 6-21–6-97. McGraw-Hill, New York.
British Standards Institution (1969). BS 3907: Part 7.
Charlot, G. (1964). "Colorimetric Determination of Elements, Principles and Methods." Elsevier, New York.
Clark, L. J. (1970). *Anal. Chem.* **42** (7), 694.
Crummett, W., and Hummel, R. (1966). *Anal. Chem.* **38,** 404R.
Crummett, W., and Hummel, R. (1968). *Anal. Chem.* **40,** 330R.
Crummett, W., and Hummel, R. (1970). *Anal. Chem.* **42,** 239R.
Danchik, R. S., and Boltz, D. F. (1968). *Anal. Chem.* **40,** 2215.
Danchik, R. S., and Oliver, R. T. (1970). *Anal. Chem.* **42,** 798.
Furman, N. H. (ed.) (1962). "Scott's Standard Methods of Chemical Analysis," 6th ed., Vols. I and II. Van Nostrand Reinhold, Princeton, New Jersey.
Goldringer, L. S. R., Hawes, R. C., Hare, G. H., Beckman, A. O., and Stickney, M. E. (1953). *Anal. Chem.* **25,** 869.
Grinstead, R. R., and Snider, S. (1967). *Analyst* **92,** 532.
Gupta, H. K. L., and Boltz, D. F. (1971). *Anal. Lett.* **4** (3), 161.
Hargis, L. G., and Boltz, D. F. (1965). *Anal. Chem.* **37,** 240.
Headridge, J. B. (1961). "Photometric Titrations." Pergamon, Oxford.
Hershenson, H. M. (1956). "Ultraviolet and Visible Spectra," Index for 1930–1954; (1961). Index for 1955–1959. Academic Press, New York.
Howell, J. A., Linington, G. E., and Boltz, D. F. (1970). *Microchem. J.* **15,** 598.
Jacobson, E. F. (1968). *U.S. Govt. Res. Develop. Rep.* **68** (8), 80.
Job, P. (1928). *Ann. Chim.* (*Paris*) **9,** 113.
Kamlet, M. J. (ed.) (1960). "Organic Electronic Spectral Data," Vol. 1 (1946–1952). Wiley (Interscience), New York.
Kanie, T. (1957). *Japan Anal.* **6,** 711.
Lang, L. (1961–1965). "Absorption Spectra in the Ultraviolet and Visible Region," 5 Volumes. Academic Press, New York.
Lapid, Y., and Pickholtz, Y. (1969). *Israel J. Chem.* **7,** 159.
Likussar, W., and Boltz, D. F. (1971a). *Anal. Chem.* **43,** 1265.
Likussar, W., and Boltz, D. F. (1971b). *Anal. Chem.* **43,** 1273.
Likussar, W., Sagan, C., and Boltz, D. F. (1970a). *Mikrochim. Acta* **4,** 683.
Likussar, W., Sparks, G. E., and Boltz, D. F. (1970b). *Anal. Chim. Acta* **52,** 349.
Lueck, C. H., and Boltz, D. F. (1956). *Anal. Chem.* **28,** 1168.
Megregian, S. (1954). *Anal. Chem.* **26,** 1161.
Miranda, C., and Conte, P. (1971). *Appl. Spectrosc.* **25,** 557.
Mustafin, I. S., Molot, L. A., and Arkhangel'skaya, A. S. (1967). *Zh. Anal. Khim.* **22,** 1808.

Pakalns, P., and Flynn, W. W. (1967). *Anal. Chim. Acta* **38,** 403.
Parker, G. A., and Boltz, D. F. (1968). *Anal. Lett.* **1** (11), 679.
Ringbom, A. (1963). "Complexation in Analytical Chemistry." Wiley (Interscience), New York.
Ringbom, A. (1939). *Z. Anal. Chem.* **115,** 132.
Sandell, E. B. (1959). "Colorimetric Determination of Traces of Metals." Wiley (Interscience), New York.
Sadtler Res. Lab., Ultraviolet Reference Spectra. Philadelphia, Pennsylvania.
Scott, A. L. (1964). "Interpretation of the Ultraviolet Spectra of Natural Products." Macmillan, New York.
Shibata, S., Uchiumi, A., Sasaki, S., and Goto, K. (1969). *Anal. Chim. Acta* **44,** 345.
Silverstein, R. M., and Bassler, G. L. (1967). "Spectrometric Identification of Organic Compounds," 2nd ed. Wiley, New York.
Slavin, W. (1963). *Anal. Chem.* **35,** 561.
Snell, F. D., and Snell, C. T. (1959). "Colorimetric Methods of Analysis," 3rd ed., Vols. II and IIA. Van Nostrand Reinhold, Princeton, New Jersey.
Snell, F. D., and Snell, C. T. (1967). "Colorimetric Methods of Analysis," 3rd ed., Vol. IVA. Van Nostrand, Reinhold, Princeton, New Jersey.
Snell, F. D., and Snell, C. T. (1970). "Colorimetric Methods of Analysis," 3rd ed., Vol. IVAA. Van Nostrand, Reinhold, Princeton, New Jersey.
Sommer, L., and Kuban, V. (1968). *Anal. Chim. Acta* **44,** 333.
Stephens, B. G., and Suddeth, H. A. (1967). *Anal. Chem.* **39,** 1478.
Tolk, A., Tap, W. A., and Lingerak, W. A. (1969). *Talanta* **16,** 111.
Ungnade, H. E. (1960). "Organic Electronic Spectral Data," Vol. 2 (1953–1955) Wiley (Interscience), New York.
Wawrzyczek, W., and Polak, K. (1967). *Z. Anal. Chem.* **228,** 433.
Weiss, K. G., and Boltz, D. F. (1971). *Anal. Chim. Acta* **55,** 77.
Ziegler, M., and Ziegeler, L. (1967). *Talanta* **14,** 1121.

Selected Reading

Boltz, D. F. (1952). "Selected Topics in Modern Instrumental Analysis," p. 105. Prentice-Hall, New York.
Boltz, D. F. (1966). "Standard Methods of Chemical Analysis," 6th ed. (F. J. Welcher, ed.), Vol. 3, Pt. A, Ch. 2. Van Nostrand, Princeton, New Jersey.
Boltz, D. F., and Gupta, H. K. L. (1971). "Vistas in Analytical Chemistry" (M. N. Sastri, ed.), Ch. 1. S. Chand, New Delhi.
Boltz, D. F., and Mellon, M. G. (1966). "Standard Methods of Chemical Analysis," 6th ed. (F. J. Welcher, ed.), Vol. 3, Pt. A, Ch. 1 Van Nostrand, Princeton, New Jersey.
Fritz, J. S., and Hammond, G. S. (1957). "Quantitative Organic Analysis." Wiley, New York.
Kharlamov, I. P. (1969). "Spectrophotometric Analysis of Alloys," *Metallurgy* Publishing House, Moscow.
Meehan, E. J. (1964). "Optical Methods of Analysis." Wiley (Interscience), New York.
Mellon, M. G. (1950). "Analytical Absorption Spectroscopy." Wiley, New York.
Parker, G. A., and Boltz, D. F. (1971). "Modern Methods of Geochemical Analysis" (R. E. Wainerdi and E. A. Uken, eds.), Ch. 5. Plenum, New York.
Willard, H. H., Merritt, L. L., Jr., and Dean, J. A. (1965). "Instrumental Methods of Analysis," 4th ed. Van Nostrand Company, Princeton, New Jersey.

CHAPTER 19

X-Ray Photoelectron Spectrometry (ESCA)

Warren G. Proctor

Varian Associates
Palo Alto, California

Introduction

ESCA in a Few Words

X-ray photoelectron spectroscopy calls for the exposure of a sample of interest to a source of monochromatic x rays, followed subsequently by analysis of the energy of the spectrum of emitted photoelectrons from the sample. One measures thereby the binding energies of the electrons in both the deep-lying shells and valence bands, providing in turn information of a chemical nature about the bulk sample, but in particular about its surface.

The experiment must, of course, be performed in a vacuum. Samples are generally solids—small plates, powders, or frozen liquids.

A Little History

DeBroglie (1921) and Robinson (1923), during the 1920s, studied photographically the distribution of kinetic energies of electrons emitted by samples when exposed to x rays, finding sharp edges where the photoabsorption suddenly increased. The practical development of the technique, however, had to wait until the startling discovery in 1954 by K. Siegbahn that photoelectrons were emitted from the sample unattenuated in energy by any scattering whatsoever, resulting in the observation of a sharp line at the foot of the emission edges observed earlier. A few years later the first chemical shift was reported where copper was used as the specimen. Approximately a decade of pioneering work by Siegbahn and his collaborators was necessary before the technique received the attention of the audience that now uses it. These early investigations and the subsequent development of ESCA into a new and unique form of spectroscopy are summarized in an important book written by Siegbahn and his collaborators (1967) with the title *Electron Spectroscopy for Chemical Analysis*, from which the name ESCA has been taken.

1 Instrumentation and Some Applications

1.1 Instrumentation

Figure 1 sketches the cross section of the x-ray tube and electron optical system of a photoelectron spectrometer which has been constructed with cylindrical symmetry, about the center line from the sample to the electron multiplier. An electron current, measuring about 100 mA, is accelerated from a tungsten filament toward the water cooled anode; the accelerating potential, marked H.V., is about 10 kV. The anode, which then must

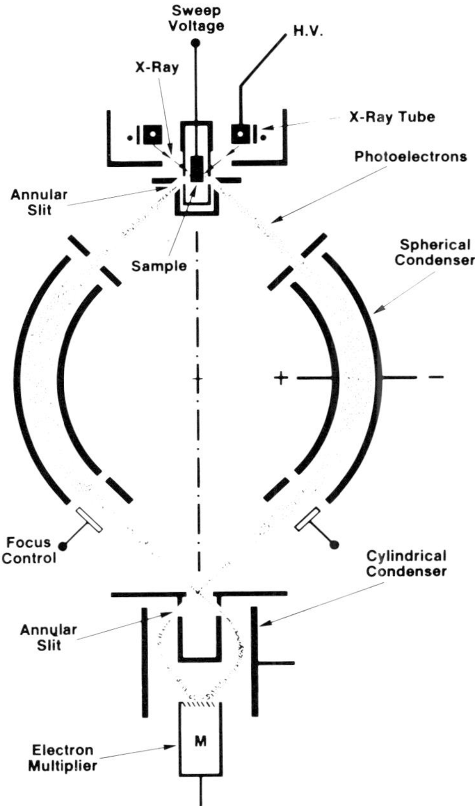

FIG. 1. Cross section of an x-ray photoelectron spectrometer.

dissipate about 1 kW, is constructed of copper. A layer of a second metal, magnesium for example, deposited on the anode produces the characteristic x rays. This classical x-ray generator provides a very weak continuum of x rays, dominated by the intense but narrow $K_{\alpha_{1,2}}$ line. For aluminum, this line has an energy of 1456.6 eV and for magnesium 1253.6 eV.

Unless effort is made to monochromatize the x rays, it is important to use a characteristic radiation that is as narrow as possible because the width of the subsequent photoelectron line cannot be narrower than that of the impinging x radiation. Magnesium and aluminum have been found to have characteristic radiations of satisfactory width—about 1-eV full-width at half-maximum. An examination of an atomic energy level chart, reproduced in part in Table 1, will show that what is called the characteristic radiation

TABLE 1

TABLE OF BINDING ENERGIES (eV) OF THE ATOMIC LEVELS OF THE FIRST 45 ELEMENTS[a]

<table>
<tr><th></th><th>$1s_{1/2}$
K</th><th>$2s_{1/2}$
L_I</th><th>$2p_{1/2}$
L_{II}</th><th>$2p_{3/2}$
L_{III}</th><th>$3s_{1/2}$
M_I</th><th>$3p_{1/2}$
M_{II}</th><th>$3p_{3/2}$
M_{III}</th><th>$3d_{3/2}$
M_{IV}</th><th>$3d_{5/2}$
M_V</th><th>$4s_{1/2}$
N_I</th><th>$4p_{1/2}$
N_{II}</th><th>$4p_{3/2}$
N_{III}</th><th>$4d_{3/2}$
N_{IV}</th><th>$4d_{5/2}$
N_V</th></tr>
<tr><td>1 H</td><td>14</td><td></td><td></td><td></td><td></td><td></td><td></td><td></td><td></td><td></td><td></td><td></td><td></td><td></td></tr>
<tr><td>2 He</td><td>25</td><td></td><td></td><td></td><td></td><td></td><td></td><td></td><td></td><td></td><td></td><td></td><td></td><td></td></tr>
<tr><td>3 Li</td><td>55</td><td></td><td></td><td></td><td></td><td></td><td></td><td></td><td></td><td></td><td></td><td></td><td></td><td></td></tr>
<tr><td>4 Be</td><td>111</td><td></td><td></td><td></td><td></td><td></td><td></td><td></td><td></td><td></td><td></td><td></td><td></td><td></td></tr>
<tr><td>5 B</td><td>188</td><td></td><td colspan="2">5</td><td></td><td></td><td></td><td></td><td></td><td></td><td></td><td></td><td></td><td></td></tr>
<tr><td>6 C</td><td>284</td><td></td><td colspan="2">7</td><td></td><td></td><td></td><td></td><td></td><td></td><td></td><td></td><td></td><td></td></tr>
<tr><td>7 N</td><td>399</td><td></td><td colspan="2">9</td><td></td><td></td><td></td><td></td><td></td><td></td><td></td><td></td><td></td><td></td></tr>
<tr><td>8 O</td><td>532</td><td>24</td><td colspan="2">7</td><td></td><td></td><td></td><td></td><td></td><td></td><td></td><td></td><td></td><td></td></tr>
<tr><td>9 F</td><td>686</td><td>31</td><td colspan="2">9</td><td></td><td></td><td></td><td></td><td></td><td></td><td></td><td></td><td></td><td></td></tr>
<tr><td>10 Ne</td><td>867</td><td>45</td><td colspan="2">18</td><td></td><td></td><td></td><td></td><td></td><td></td><td></td><td></td><td></td><td></td></tr>
<tr><td>11 Na</td><td>1072</td><td>63</td><td colspan="2">31</td><td>1</td><td></td><td></td><td></td><td></td><td></td><td></td><td></td><td></td><td></td></tr>
<tr><td>12 Mg</td><td>1305</td><td>89</td><td colspan="2">52</td><td>2</td><td></td><td></td><td></td><td></td><td></td><td></td><td></td><td></td><td></td></tr>
<tr><td>13 Al</td><td></td><td>118</td><td>74</td><td>73</td><td>1</td><td></td><td></td><td></td><td></td><td></td><td></td><td></td><td></td><td></td></tr>
<tr><td>14 Si</td><td></td><td>149</td><td>100</td><td>99</td><td>8</td><td colspan="2">3</td><td></td><td></td><td></td><td></td><td></td><td></td><td></td></tr>
<tr><td>15 P</td><td></td><td>189</td><td>136</td><td>135</td><td>16</td><td colspan="2">10</td><td></td><td></td><td></td><td></td><td></td><td></td><td></td></tr>
<tr><td>16 S</td><td></td><td>229</td><td>165</td><td>164</td><td>16</td><td colspan="2">8</td><td></td><td></td><td></td><td></td><td></td><td></td><td></td></tr>
<tr><td>17 Cl</td><td></td><td>270</td><td>202</td><td>200</td><td>18</td><td colspan="2">7</td><td></td><td></td><td></td><td></td><td></td><td></td><td></td></tr>
<tr><td>18 A</td><td></td><td>320</td><td>247</td><td>245</td><td>25</td><td colspan="2">12</td><td></td><td></td><td></td><td></td><td></td><td></td><td></td></tr>
<tr><td>19 K</td><td></td><td>377</td><td>297</td><td>294</td><td>34</td><td colspan="2">18</td><td></td><td></td><td></td><td></td><td></td><td></td><td></td></tr>
<tr><td>20 Ca</td><td></td><td>438</td><td>350</td><td>347</td><td>44</td><td colspan="2">26</td><td colspan="2">5</td><td></td><td></td><td></td><td></td><td></td></tr>
</table>

21 Sc	500	407	402	54	32		7					
22 Ti	564	461	455	59	34		3					
23 V	628	520	513	66	38		2					
24 Cr	695	584	575	74	43		2					
25 Mn	769	652	641	84	49		4					
26 Fe	846	723	710	95	56		6					
27 Co	926	794	779	101	60		3					
28 Ni	1008	872	855	112	68		4					
29 Cu	1096	951	931	120	74		2					
30 Zn	1194	1044	1021	137	87		9					
31 Ga	1298	1143	1116	158	107	103	18			1		
32 Ge	1413	1249	1217	181	129	122	29			3		
33 As		1359	1323	204	147	141	41			3		
34 Se		1476	1436	232	168	162	57			6		
35 Br				257	189	182	70	69	27	5		
36 Kr				289	223	214	89		24	11		
37 Rb				322	248	239	112	111	30	15	14	
38 Sr				358	280	269	135	133	38	20		
39 Y				395	313	301	160	158	46	26		3
40 Zr				431	345	331	183	180	52	29		3
41 Nb				469	379	363	208	205	58	34		4
42 Mo				505	410	393	230	227	62	35		2
43 Te				544	445	425	257	253	68	39		2
44 Ru				585	483	461	284	279	75	43		2
45 Rh				627	521	496	312	307	81	48		3

[a] Taken from K. Siegbahn *et al.* (1967), "Electron Spectroscopy for Chemical Analysis," Almquist and Wiksells, Uppsala. Levels deeper than 1450 eV have been omitted since it would be impossible to observe them when using Al or Mg as an anode.

from Mg or Al is in fact two unresolved lines; these result from the $^2p_{1/2}$ and $^2p_{3/2}$ to the 1s level transitions.

Turning again to Fig. 1, the x rays enter the sample cage through a thin aluminum window. The window completes the cage which defines the potential gradient around the sample. It also serves as a crude filter, helping to reduce further the intensity of the continuous x-ray background from the anode. This occurs because the Al $K_{\alpha_{1,2}}$ radiation is found in an absorption notch as one plots the x-ray absorption of aluminum against energy. On the higher energy side, there is a strong absorption for x rays which have enough energy to remove a K electron from the atom (1560 eV), and in the low-energy direction there is a general rise of absorption with decreasing energy. Finally, the window serves one further purpose: the x rays produce photoelectrons in the aluminum window that provide an electron current to the sample. This current serves to neutralize the surface potential of nonconducting samples which become charged from the loss of photoelectrons.

The x rays impinge upon the sample surface, and photoelectrons are ejected from all energy levels of all the atoms in the sample, including the valence and conduction bands. Thus coming through the upper annular slit are electrons which compose a spectrum of emitted energies to be measured and recorded.

The analyzer shown here is composed of portions of two concentric spheres, followed by a cylindrical condenser. The deflection voltage across the analyzing electrodes (inner and outer electrodes of the spherical condenser) may of course be so chosen as to permit electrons in a narrow range of energies to pass to the electron multiplier; the counting rate is then recorded as a function of this deflection voltage. If digital techniques are employed, the counts will be registered in a given channel, which has been selected by the sweep voltage, the contents of all channels being plotted subsequently to produce a spectrum. If analog methods are used, an average current is plotted as the spectrum is taken.

The electron current flowing through the slit of a spectrometer will be the product of the brightness B of the electron illumination of the slit, its area A, and the solid angle G of the spectrometer as viewed from the slit. B will be in units of electron current per unit area per unit solid angle. The slit area A will be proportional to the length of the slit; hence a spectrometer with an annular slit and cylindrical optics should provide a sensitivity greater than one would expect from a linear slit and sector optics, by the ratio of 360° to the width of the sector.

Figure 1 does not make clear an innovation introduced by Helmer and Weichert (1968) which is now general practice in the design of photoelectron spectrometers. Helmer and Weichert point out from simple electron optical considerations that the product of AG is proportional to R^2 $(\Delta E/E)^2$, where R is the radius of curvature of the beam center line, ΔE the line width, and E the energy of the electrons being analyzed. Thus if one were to retard the photoelectrons from the sample, shifting their initial energy E_0 to an energy E, a factor of $(E_0/E)^2$ in intensity will be gained. For example, the kinetic energy of the photoelectrons from the C 1_s level is 970 eV; retarding these before analysis to 100 eV would enhance the product AG by a factor of almost 100.

The brightness B, however, is proportional to the inverse of the analyzing energy. The reason for this is easily understood: While the velocity of the electrons toward the slit is reduced, the component of the velocity perpendicular to this direction are unaffected by the retardation; hence the beam is dispersed. The net effect of retardation is then to give an enhancement of sensitivity which is linear in $1/E$, so that, in our example, approximately an order of magnitude of sensitivity is gained by retardation.

A further advantage of retardation is that the instrumental resolution, which is a function of the energy of the electrons being analyzed, remains constant over the spectrum.

The counting rates of photoelectric lines in modern ESCA instruments are about three orders of magnitude greater than those obtained from the instruments used to provide the illustrations in the classic book, *Electron Spectroscopy for Chemical Analysis* (Siegbahn *et al.*, 1967). Indeed, an illustration in that book of the $4f_{7/2}$ emission lines of gold shows a counting rate of 20,000 electrons per minute. Today the average commercial spectrometer provides the same number of counts at the same resolution within a fraction of a second. It is not, however, the spectacular counting rates for the photoelectric emission lines of the gold 4f doublet or from the C 1s level in graphite which is important—a high counting rate for those elements insures that a reasonable counting rate for trace atoms in the sample will be obtained. Such counting rates reduce to minutes the data collection time for atoms stoichiometrically present in small molecules. For minor or semitrace elements or atoms present in small quantity in large molecules (iron in hemoglobin or cobalt in vitamin B_{12}) the collection times can approach an hour. These practical exposure times further enhance the attractiveness of a technique which, unless the sample is susceptible to radiation damage, can be regarded as nondestructive.

The availability of inexpensive small computers (16-bit word length, 8000 words) as well as the availability of digitally controlled high-precision power supplies, makes it possible to design an x-ray photoelectron spectrometer which is entirely controlled through the keyboard of the computer's teletype. Certain obvious functions must be excluded, of course—the loading of the sample, turning of vacuum valves, and the like. One may preselect a binding energy region to be studied, specifying the beginning and ending energies, the number of repetitive scans, the number of channels into which the region is divided, and the time per scan. Generally, sufficient memory will be available such that about ten such regions may be chosen, or alternatively, one large scan, using all the available memory channels. The operator is freed from manipulation of the spectrometer controls during the uneventful data gathering time. At the end of this period, the operator will find all data stored for ready presentation upon an oscilloscope or a recorder. In the latter case, alternate forms of presentation will be available: points, representing the contents of each channel in the region being studied, or a continuous line, representing a simple or weighted moving average of a number of adjacent channels. The specified experimental conditions are typed out on command and may be attached to the spectrum obtained. Data manipulative programs, generally requiring access to supplementary memory, such as would be provided by a tape deck, can provide deconvolution of a line with structure into the simple lines which compose it or elaborate filtering techniques, such as are available from the Fourier transform of the spectrum, which may reduce noise or artificially narrow the line width.

Thus, a phenomenon and an apparatus have been sketched which will produce, measure, and record photoelectrons ejected by monochromatic x rays from a sample. The method does not compete in sensitivity with x-ray fluorescence or atomic absorption for detecting the presence of elements in the sample. ESCA has another more important application: it has been found that the binding energy of the inner core levels are sensitive to the behavior of the valence electrons. It is the displacement or multiplicity of the lines observed that is of interest. The demonstration that useful chemical information can be obtained from a small amount of a solid sample has made the method one of great interest today.

1.2 An ESCA Spectrum

Figure 2 illustrates a spectrum obtained by a spectrometer of the design sketched in Fig. 1, and which uses a retarding voltage and digital techniques for storing experimental results. The sample material is NH_4NO_3, and the spectrum is that of the 1s level of nitrogen. This compound has been chosen for illustration for historical reasons: it is this very same compound with which the "chemical shift" of nuclear magnetic resonance (NMR) was discovered. Indeed, the ratio of the displacement of the nitrogen peaks to the width of the lines bears a close resemblance to those early NMR signals.

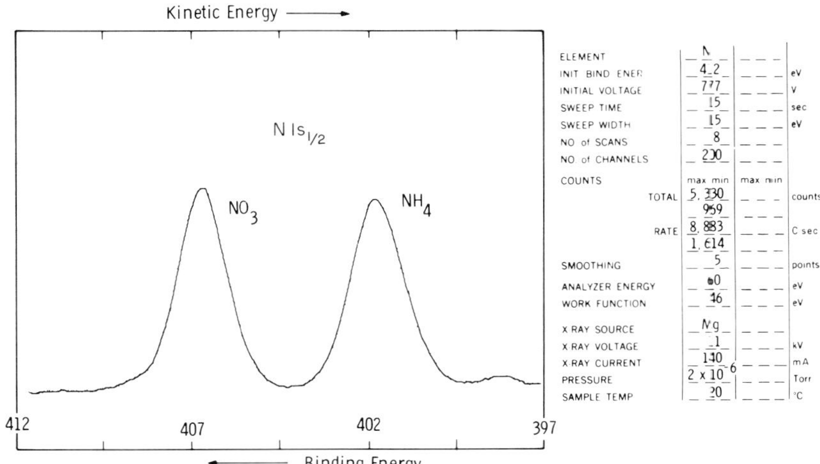

FIG. 2. X-ray photoelectron spectrum of NH_4NO_3.

The line on the higher binding energy side (~407 eV) is due to photoelectrons from the $1s_{1/2}$ level of nitrogen in the nitrate group, and that on the lower (~402 eV) from the ammonium group. The photoelectrons from the nitrogen in the nitrate group are emitted with less kinetic energy because this nitrogen, surrounded by electronegative oxygens, has lost some electronic charge, and the $1s_{1/2}$ level has been lowered by 5 eV as a consequence.

To obtain an idea of the experimental conditions used to make this spectrum, let us look at some of the parameters listed in the table contained in Fig. 2. The initial binding energy is the binding energy at the beginning of the scan, at the left-hand side. Thus the scan starts at 412-eV binding energy and the 15-V sweep width extends it down to 397 eV, in accordance with the convention which plots kinetic energy as increasing to the right, thus binding energy as decreasing to the right. To scan the 15-V scan once 15 sec were used; 8 scans were made, making a total time of observation for this compound of 120 sec, or 2 min, a fairly typical exposure. The information was collected and stored in 200 channels, and the content of each channel was subsequently plotted as a point. A line was drawn through these points which represents a moving average of 5 points (smoothing, 5 points). The actual number of counts in two of the channels is recorded: that channel with the largest number of counts (max) and that channel with the least (min). Thus 5330 counts were obtained in one of the channels near the NO_3^- peak and 969 counts in a channel along base line somewhere.

The rate presents the same information, but divided by the time spent on each channel, and thus is the counting rate of photoelectrons into each of these two channels. From the information given above, the reader may quickly check that 8883 and 1615 counts/sec are indeed those observed rates.

This line width is a convolution of the natural line width of the magnesium $K_{\alpha_{1,2}}$ radiation (about 0.8 eV), the aberrations of the analyzer, and the natural line width of these levels. The weak emission noted near 398 eV is not due to the presence of a nitrogen of an unusual binding energy, but due to a magnesium satellite radiation, the Mg $K_{\alpha_{3,4}}$ about 9 eV higher in energy. Thus this line represents again emission from the nitrate nitrogen. Because of their anticipated position and intensity, the presence of weak satellite lines noted on the lower binding energy side of strong emission lines causes the experienced observer no practical difficulty.

1.3 Another Example

Figure 3 is an example of an investigation which illustrates a good application of the technique. An examination of the 4f doublet of gold (center scan in Fig. 3) was presumed to be Au_2O_3. However two chemically different species of gold apparently exist in the sample. For comparison, this doublet, from metallic gold, is shown in the scan on the left-hand side. (Note the change of energy scale.) For calibration of the charging

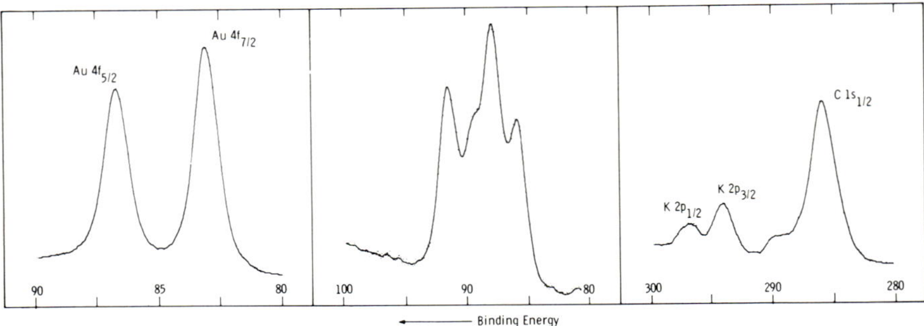

Fig. 3. Results of an examination of a product believed to be Au_2O_3. The chart on the left shows the 4f doublet of gold; the center chart is the spectrum obtained from the product for this same region; the chart on the right shows the emission spectra of K and C from the same product.

shift, a scan of the carbon was made which revealed the presence of potassium. KOH had been used to prepare the oxide from the chloride. Correction for the charging shift reveals that the sample is composed partially of the desired Au_2O_3 and partially of the entirely reduced gold itself, not to mention small amounts of K and Cl. The Cl emission lines are not shown in Fig. 3.

2 Spectral Details

2.1 The Chemical Shift

Chemical shift is the name which is given to the observed changes in the binding energies of core levels with changes in chemical bonding. This is analogous to the chemical shift dependence of the NMR frequencies in chemical compounds. It is observed that bonding which removes valence electron density from an atom will increase the observed binding energy of a core level. The variations in binding energy are not trivial, being about 10 eV for many elements. Assuming that one may reliably measure the position of an emission line to about 0.1 V, ESCA is thus a spectroscopic technique in which the ratio of full spectrum width to discernible line position is about two orders of magnitude.

One is tempted to anticipate a strong correlation between core level binding energy shifts and the chemical shifts of NMR. However, the diamagnetic term is the only mechanism contributing to the NMR chemical shift which is related to core binding energy shifts. Magnetic anisotropy of substituents and delocalized π electron systems as well as the paramagnetic term make strong if not dominant contributions to the chemical shifts of NMR. For example, the ^{13}C NMR chemical shifts of cyclohexane and benzene differ by more than 100 ppm while the C 1s binding energies are essentially identical. In those systems where the diamagnetic term can be expected to be dominant, such as the halomethanes, one obtains a good correlation of the C 1s binding energy shifts with the NMR chemical shift, as well as with the NQR coupling constant of chlorine as this atom is progressively substituted for hydrogen.

Removal of valence electron density from an atom deepens the potential well in which the inner core electrons move so that binding energies increase. *Ab initio* calculations or approximative quantum mechanical techniques are used to predict charge distributions in molecules and core energy levels; however, a simple electrostatic theory can account for main features of the observed chemical shifts. The chemical shift, in units of electron volts, is given by $\Delta\phi$ where ϕ is the binding energy of a given level for the neutral atom.

$$\Delta\phi = (\Delta q/r) + \sum_i (\Delta q_i/R_i) \tag{1}$$

Here Δq is the change in the net charge on the atom being observed, and r is its radius. Δq_i is the change in charge on the ith neighbor, and R_i is the internuclear distance to the ith atom. $\Delta q/r$ will be recognized as the change in potential of the atom under consideration; $\Delta q_i/R_i$ is the potential at the atom under consideration due to a change Δq_i a distance R_i away. Thus as a photoelectron is ejected, it is attracted or repelled by the sphere from which it comes and whose potential now differs from zero by $\Delta\phi$.

This explanation provides a phenomenologically satisfying picture of the direct role of the electronegativity of neighboring atoms in producing the chemical shifts. The second term in Eq. (1) brings to attention that, in a solid, the chemical shift of a given atom also depends upon a summation over all atoms in the sample. This is known as the Madelung potential.

As Citrin *et al.* (1972) have shown, the Madelung sum is far from negligible. There is, for example, a 9.3-eV increase in the binding energy of the 2p level of Na as it ionized to Na^+; however there is an 8.9-eV decrease in the binding energy of that state as the ion is placed into the ionic crystal NaCl. The same authors have shown that the simple electrostatic theory sketched here can account for the relative binding energies of different species of alkali halide crystals to within 0.1 eV.

Chemical shifts cannot be revealed by x-ray fluorescence spectrometry because they have, to a first approximation, the same magnitude for K and L levels. The situation is too complex to be useful for Auger spectrometry, where three levels are involved.

2.2 Auger Lines

An atom, having lost a K electron by photoemission, generally returns to its ground state by the classical sequence of electronic transitions during which photons are radiated. However, the atom may, as one step in its return to the ground state, eject instead another electron—the well known Auger process, and one which is indeed more probable than radiative transitions for light elements. The kinetic energy of this electron will depend upon the three states which are involved in the process. Let us assume that an electron falls from the L_I level to the vacancy in the K shell, while a second electron, the Auger electron, is ejected from the L_{II} shell. The Auger electron will then be identified as a KL_IL_{II} electron, and its kinetic energy will be approximately equal to the difference between energies of the K and L_I levels minus the binding energy of the L_{II} level for the atom in which one K electron is missing. Auger energies are generally known from experimental measurements and are tabulated or plotted.

Note that the Auger energy is a property of the atom in question and does not depend upon the energy of the x ray used to excite the atom. Although ESCA spectrometers to indeed measure electron kinetic energies, it is convenient to present the emission lines on a scale of binding energies. Thus an Auger line will appear on an ESCA chart as a line with a certain binding energy. If it is suspected that the line may be an Auger line, its apparent binding energy is subtracted from the known x-ray energy used in the spectrometer to obtain its kinetic energy; this value is then compared with the published chart or table. In the case of an ambiguity where the line may be either an Auger line or a photoelectric emission line, positive identification as an Auger line may only be made by changing the x-ray energy; this will obviously displace an Auger line on the chart of binding energies, but will not displace a photoemission line.

A useful chart of KLL Auger energies is found in Appendix 4 of Siegbahn *et al.* (1967). Since the x-ray energy for the Mg K_α radiation is 1253.6 eV, all photoelectrons must have less than this energy; hence KLL Auger electrons from elements with Z equal to or less than 12 will be observable. As a practical matter, Auger electrons from C, O, and Na are commonly observed in ESCA spectrometry. An examination of Auger spectra in x-ray photoelectron spectrometry has been given by Wagner (1972).

2.3 Shake-up and Shake-off Processes

Shake-up and shake-off are names which have found their way into common usage for processes in which two electrons are photoexcited simultaneously. The former describes the promotion of an electron, commonly from a valence band to the top of the conduction band, that is, from one well-defined energy level in the sample to another, at the same time as the ejection of the photoelectron. If the energy difference between these two levels is eV', then it is clear that the process will give rise to a subsidiary peak V' volts on the higher binding energy side of the principle photoelectric line. If a variety of inner excitations occur, then a number of lines will be seen. Novokov and Prins (1972) point out the excellent agreement between the displacement of peaks and published ultraviolet absorption spectra for NiO.

The shake-off process is one in which both electrons are photoexcited to the continuum. In this case, since the energy of the second photoelectron is not precisely defined, a distinct satellite will not be produced, although the emission edge, at which the second photoelectron enters the continuum at zero energy, can be clearly seen. The shake-off spectrum found on the higher binding energy side of the $ls_{1/2}$ level of gaseous neon showed that this process, at least for this one element, was not as rare as one would otherwise suspect: about 18% of the photoabsorptions were double electron events, and in 85% of these, both electrons went into the continuum.

These phenomena, presently regarded as somewhat of a nuisance, will no doubt provide a new tool for the investigation of the overlap of valence orbitals with core vacancies, for which little information is presently available. For example, Gelius *et al.* (1970) have studied the linear molecule C_3O_2 and demonstrated that no shake-up electrons are emitted from the central atom, which can be identified by its chemical shift. Evidently valence electrons on the outer carbons have a much greater overlap with the K shell than does the inner one.

An even more striking example has been provided by Novokov and Prins (1972) who demonstrated that adsorbed oxygen (or possibly water vapor) on the surface of compounds such as Cu_2O, NiO, and a number of others, was essential for the production of shake-off electrons.

Additional phenomena, such as double ionization, i.e., shake-off accompanied by shake-up, have been proposed. A reader particularly interested in these effects is referred to the review of the subject provided by Allan and Siegbahn (1971).

2.4 Configuration Interaction

There is another process which also provides satellite lines on the high binding energy side of the main photoelectric line and which may be confused with the shake-up process: the generation of a satellite which is a consequence of configuration interaction. This effect occurs whenever there is a simultaneous excitation of one or more electrons in the residual atom to a final state of the same angular momentum and parity as that in which the atom would otherwise have been left. Configuration interaction has been described by Wertheim and Rosencwaig (1971) who observed these satellites in a number alkali halides, pointing out that transition elements, where unpaired d or f electrons may recombine to new states differing only in energy, will be a likely source for satellites as well.

2.5 Plasmon Excitation

When studying the x-ray photoelectron spectra of metallic magnesium, we see two lines, displaced by 7.1 and 10.6 eV and having an intensity of about 10% of that of the principle line, on the higher binding energy side of the principle photoelectric line. These lines are a consequence of the excitation by a photoelectron of a quantized electron plasma oscillation (plasmon) of which there are two types, one characteristic of the bulk and one characteristic of the surface. These peaks, then, represent observation of photoelectrons which have suffered no other energy losses other than the excitation of these plasmons. Plasmon oscillations are possible in dielectrics but as a practical matter in photoelectron spectroscopy they are seen only when fairly clean metals are studied.

2.6 Charging Effect

A meaningful determination of a chemical shift depends upon measuring the absolute binding energy of an atomic level. Unfortunately it is not possible with a nonconducting sample simply to use observed binding energy values because the surface may have become charged. The charging is a consequence of the electronic charge lost in the photoemission process and the charge gained from electrons generally as a result of photoemission elsewhere in the sample cavity, but particularly in the x-ray window. The spectrometer simply measures the kinetic energy of the photoelectron; it cannot distinguish between a loss of kinetic energy due to a higher core binding energy or to a more positive surface charge.

Because of the lack of a convincing demonstration that any other method is much better, most experimenters continue to calibrate surface charging by measuring the displacement of the C 1s line of the ever present carbon impurity relative to the absolute calibration of the spectrometer. The method is simple in that no preparation is necessary, and the experimenter is comforted by the knowledge that he is in good company in choosing it, for this is the method adopted by Siegbahn and many others.

Hnatowich (1971) recommends the evaporation of a light layer of a noble metal on the sample surface; the charging is measured by noting the difference in binding energy of a line observed from islands of the metal on the sample surface and islands found in electrical contact with the spectrometer. Bremser and Linneman (1971) discuss briefly various methods and conclude that the best method is to measure line positions relative to the F 1s emission line obtained by grinding a small amount of LiF into one's sample.

In some experiments, where one is comparing an element in one molecule to the same element in another, one can reliably use as an internal standard an atom contained in both molecules, which one knows will remain chemically unchanged.

2.7 Satellite Lines

Figure 4 shows the result of an examination of a titanium plate of uncertain history. Other examples of the weak satellite emission line already noted at 398 eV in Fig. 2, a consequence of the Mg $K_{\alpha_{3,4}}$ x ray, may be seen. In this figure, a "survey scan" from 600 eV down to 0 eV binding energy was made in order to get an impression of the main photoelectric lines which would be easily observable for more detailed study. Although the data in the survey scan could have been presented in one continuous scan, obviously more detail is available to the observer if smaller regions may be independently amplified and plotted. In the scan from 300- to 200-eV binding energy, where only photoelectrons from the C 1s level of carbon are observed, the gain has been increased to such an extent that the principle C 1s photoelectric line could not be contained on the chart in order to bring out the weaker satellite lines.

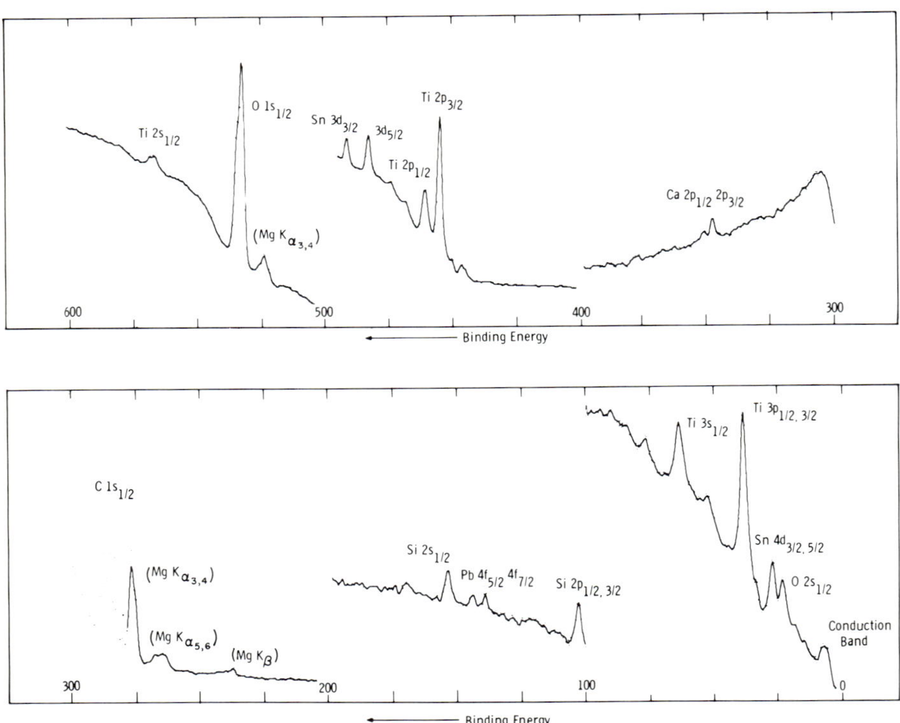

Fig. 4. The x-ray photoelectron spectrum from a sample of metallic titanium.

The $K_{\alpha_{3,4}}$ x-ray transition, like the $K_{\alpha_{1,2}}$ transition, is one from the unresolved 2p states to the 1s vacancy, but in an atom which has suffered the additional excitation of a p electron to the valence band. A carefully performed experiment will reveal another emission line between the $K_{\alpha_{3,4}}$ and $K_{\alpha_{1,2}}$ lines, the $K_{\alpha'}$ lines. The $K_{\alpha_{5,6}}$ line, located about 20 V lower than the $K_{\alpha_{1,2}}$, and the K_β line, located about 50 eV lower than the $K_{\alpha_{1,2}}$, each have an intensity of about 10% of the $K_{\alpha_{3,4}}$ and may be seen in this scan. The $K_{\alpha_{5,6}}$, like the $K_{\alpha_{3,4}}$, is a consequence of another excitation, while the K_β is due to a transition directly from the valence band to the 1s vacancy.

In practice, only photoemission lines from the $K_{\alpha_{3,4}}$ are seen, and then following strong photoelectric lines, such as O $1s_{1/2}$ and Ti 2p lines in Fig. 4. The weaker satellites are generally never observed.

2.8 Anisotropic Effects

The variation of intensity of photoelectron emission with crystal-line direction has been observed. The first experiment, by Siegbahn and co-workers (1970), was made using NaCl as a sample. Photoelectrons emitted from the {100} surface were studied as a function of angle from the normal in a principle plane. Although the de Broglie wavelength is short compared to the lattice spacing for some of the photoelectron energies, the results were qualitatively explained in terms of diffraction. This result is confirmed by a similar study on a single crystal of gold reported by Fadley and Bergstrom (1972). Anisotropy of spin polarization of photoelectrons emitted from ferromagnetic materials has been observed as well (Busch *et al.*, 1972).

2.9 Some Further Details

If a wide scan is made, a spectrum resembling that of Fig. 4 is generally obtained. A striking feature of such a spectrum is the rise of the level of the base line toward higher binding energies. The reason for this is that photoelectrons, on being scattered, lose kinetic energy and thus contribute to the background level at higher apparent binding energies. Such steps in the background level are particularly noticeable after strong lines. In the close neighborhood of a line, however, an emission line, such as those illustrated in Figs. 2 and 3, the immediate background appears more or less flat. Thus it appears that photoelectrons do not lose, with high probability, small (less than about 10 eV) amounts of energy.

The reader will note the presence of a strong conduction band, which illustrates that conduction and valence bands in solids may be easily studied by this technique. One sees other peaks which do not appear to be accidents of noise; their ultimate identification will only be revealed with further work.

3 Sample Preparation

3.1 Sampling Depth

Before beginning a review of sample-handling techniques, a discussion of the sampling depth, or the depth from which the photoelectrons which one observes originated, is in order.

Siegbahn and collaborators (1967), in a series of experiments involving the deposition of monolayers of Br and I substituted stearic acid, reach the conclusion that, for photoelectrons from carbon (about 1200-eV kinetic energy in these experiments), the half-depth, that is the depth from which one-half the photoelectric current has been generated, was about 5 nm (50 Å). It was pointed out, however, that range–energy curves for electrons in the kiloelectronvolt region are approximately the same for different materials when the range is expressed in units of mass per unit area. This suggestion was used by Steinhard *et al.* (1972), who, in experiments in which carbon was evaporated onto gold, reached the conclusion that the dependence on depth of the observed photoelectrons can be given by

$$I/I_0 = 1 - e^{-\mu l} \tag{2}$$

where I is the photoelectric current to depth l and I_0 the total current. μ was determined to be $(3.2 \pm 0.1)\ 10^3/E$ (eV) μg/cm^2. Note that the assumption that the mass scattering coefficient is inversely proportional to energy has been made. Applying these equations to an organic material of unit density, assuming a Mg anode for which the kinetic energy of the C 1s electrons would be 970 eV, a half-depth of 2.1 nm (21 Å) is obtained. Examining photoelectrons generated by Mg K_α x rays from the 3p level of iron, which has a binding energy of 32 eV, half of the signal intensity is produced within the outermost 0.35 nm (3.5 Å).

3.2 Elementary Sample Handling

It is evident that ESCA is very much a surface technique and that x-ray photoelectron spectra of monolayers is entirely a practical matter. Reliable data on bulk properties will therefore depend upon the examination of clean surfaces. Fortunately most substances provide meaningful data without unusual sample-handling techniques, other than elementary common-sense precautions, such as not touching the sample once it is mounted for examination. Oxygen or moisture sensitive samples, which have been sealed in glass tubes, may be opened in a dry box, mounted onto the probe, and transferred in a portable housing under the pressure of an inert gas to the spectrometer.

3.3 Sputter Cleaning

A conducting sample may be cleaned by argon ion sputtering, by using the sample as the cathode in an argon discharge. The chamber in which the sputtering is done is filled to approximately 50 mTorr of argon, and a discharge is provided by a power supply capable of delivering about 1000 V and 20 mA. Experience has shown that 1 or 2 min of such cleaning will generally remove most traces of oxygen or oxygen bearing impurities and greatly reduce the surface carbon contaminants. Samples will be heated by the dissipation to such an extent that, unless the cleaning is limited in time, the sample may melt. For the most careful cleaning, the geometry should be such that only the sample, and none of the probe holder, serves as the cathode, for otherwise atoms sputtered from the probe holder will be found on the sample.

Figure 5 illustrates the removal of a surface oxide from the semiconductor Sb_2Se_3. Each member of the Sb 3d doublet on the right shows photoelectrons both from the surface oxide and from the bulk material, with the intensity of the $3d_{5/2}$ line somewhat enhanced by the O $1s_{1/2}$ line, which has a binding energy of 532 eV in the neutral atom. After sputtering the emission spectrum on the left is obtained; the oxide layer has been removed but a small residual O $1s_{1/2}$ line may be observed.

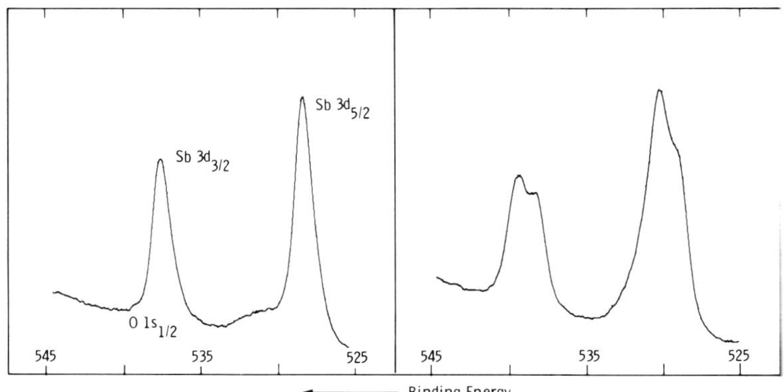

FIG. 5. The effect of argon sputtering: the sputtered (left) and the unsputtered (right) spectra of the 3d doublet region of Sb from the semiconductor Sb_2Se_3.

For cleaning purposes, nonconducting samples fall into two classes: those which may be formed into thin layers over a conductor, such as powders, and large crystals. Because a nonconducting surface will become charged at once and prevent further impact by charged ions, the former may be sputtered successfully by using a radio-frequency discharge. Unfortunately the danger of back scattering of atoms from the other electrode, generally the walls of the chamber itself, is always present.

Large nonconducting objects, such as a single crystal, may be sputtered by neutral ions which have been accelerated by momentum exchange collisions in the discharge. The crystal is suspended just below the cathode, which is surrounded by a glass cylinder but not protruding from it. Ions are accelerated in the general direction of the cathode, but must change direction immediately before the crystal in order to follow the lines of force. Collisions in this area will cause neutral argon atoms of high momentum to reach the crystal surface. Argon ion guns used with electron guns to neutralize the charge also have been utilized.

3.4 Evaporation

Volatile liquids may be studied in the frozen state if the cooled probe is exposed in a preparation chamber to the vapor generated by the liquid at room temperature.

Metals evaporated from heated wires located near the sample holder exhibit a clean surface. If an argon discharge is maintained at the same time as the evaporation, an exceptionally clean surface will be created. For materials which will sublimate, simple techniques for evaporating these materials from simple heatable cups onto the sample holder without contaminating the entire preparation chamber are easily designed.

3.5 Temperature Variation of the Sample

As in most forms of spectrometry, there is a need in ESCA to vary the temperature of the sample while it is under observation. Temperature changes will, for example, permit the study of charge redistribution with phase changes. The change of temperature of the sample will also make possible certain practical operations, such as the drying of sample materials, the freezing of liquids, and the evaporation of volatile impurities.

The temperature of a sample can only, with some reliability, be prescribed if it is a good thermal conductor and in substantial contact with the variable temperature probe. The probe itself is generally heated with a flow of hot air, a thermocouple sensing the temperature at the point as near the sample as practicable, feeding back error information such as to keep the thermocouple at the constant and preset temperature. The same is of course true for cooling, although in this case one has the additional freedom of filling the probe with liquid nitrogen or freezing mixtures to achieve certain fixed temperatures.

In a sample preparation chamber, however, samples may be subjected to extreme, unregulated temperatures by experiments of an arbitrary kind. For example, a conducting sample material may be heated, for cleaning purposes, to a temperature near its melting point by electron bombardment.

3.6 Sample Quantity and Geometry

The most likely form of samples in the solid state will be a powder. In the design described above, such powders are often dusted onto the sample holder, made adhesive by the use of commercial cellophane tape. Sufficient powder to dust several square centimeters forms the practical basis for estimating the amount of sample necessary, generally some 50 mg. Powders may also be compressed into pellets and treated as solids.

The amount of material actually providing photoelectrons is, of course, many orders of magnitude smaller. If we assume that the photoelectrons arise from a few times 1 nm (10 Å) in depth, then the sample actually being examined is only a few hundred nanograms. If the sample material is a soluble one, then a physically much smaller sample quantity may indeed be used: the sample is dissolved and brushed onto the sample holder. After evaporation such samples, which often cannot be detected by eye, provide excellent x-ray photoelectron spectra.

In sector spectrometers, or those in which the x-ray beam and the subsequent photoelectron optics may be described as lying within certain finite sectors, flat samples, such as foils, may be studied without difficulty. Indeed, this geometry has the advantage that often only one side of the sample is prepared for examination.

For the "axial" spectrometer which has been described above, the sample is irradiated on all sides. In this case, one may mask the back of the sample with a material which does not contain the same elements as the material on the face; or, one may place two such samples back to back, making allowance for photoelectrons which are generated along the edges; or, one may modify the spectrometer to accept electrons only from a 90° sector. This latter is done by either masking the x-ray window and slits, or by generating x rays from a portion of the annular anode if the filament may be heated in parts.

3.7 The ESCA of Gases

The x-ray photoelectron spectra of gases provide information on molecular levels which are broadened in the solid state. Siegbahn and his collaborators (1969) have described the ESCA of gases in detail. The freedom to examine gases lends an additional strength to an already powerful tool. Modifications can be made to the analyzer sketched in Fig. 1 in order to provide this facility. A solid baffle to isolate the anode and sample area from the analyzer is introduced and a differential pumping access is provided to the upper portion of this area. A 90° sector of the slits, located just under a 90° sector x-ray window, is also used. This geometry reduces background counts essentially to zero.

4 Concluding Remarks

4.1 An Important Future Development

Designs are actively underway at the University of Wisconsin and at Stanford University to use the bremstrahlung from the acceleration of high-energy electrons captured in electron storage rings as an x-ray source. Very high intensities will be available over an extremely broad range of energies. An energy may be selected for which ideal monochromator crystals can be chosen. The proposed spectrometer at Stanford is being designed to provide x rays with a full width at half-maximum of 0.05 V! With a source this monochromatic, one will always be observing the natural line widths of the photoelectric lines of the sample. If supported financially, these spectrometers may begin operation in 1974.

X-ray photoelectron spectroscopy is at this writing a fast-developing field. Many researchers, alert to the advantages of a more monochromatic source, are investigating natural sources of radiation in the hope of eventually finding a practical alternative to Al or Mg. One candidate, yttrium, provides a characteristic radiation of 0.4 eV but an energy too low to be of serious merit for the study of core electrons, not to mention the practical difficulties of working with this rapidly oxidizing metal. Gaseous sources of x rays may provide narrow lines, but intensities will obviously be low.

4.2 Manufacturers of Commerical ESCA Instrumentation

We conclude with a listing of known manufacturers of commercial ESCA instrumentation, in alphabetical order.

AEI Scientific Apparatus Ltd., Urmston, Manchester, England

Hewlett-Packard Corp., Palo Alto, California
MacPherson Instrument Corp., Acton, Massachusetts
Vacuum Generators Ltd., East Grinstead, Sussex, England
Varian Associates, Palo Alto, California

ESCA spectrometers are also manufactured by the following Japanese companies: Hitachi, Jasco, JEOL, Kokusai Denki, and Rigaku.

References

Allan, C. J., and Siegbahn, K. (1971). Rep. UUIP-754, Inst. of Phys., Uppsala Univ.

Bremser, W., and Linneman, F. (1971). *Chem. Z.* **95,** 1011.

Busch, G., Campagna, M., and Siegmann, H. (1972). "Electron Spectroscopy," pp. 827–834. North Holland Publ., Amsterdam.

Citrin, P., Shaw, R., Packer, A., and Thomas, T. (1972). "Electron Spectroscopy," pp. 691–706. North Holland Publ., Amsterdam.

De Broglie, M. (1921). *C. R. Acad. Sci. Paris* **172,** 274.

Fadley, C. S., and Bergstrom, S. (1972). "Electron Spectroscopy," pp. 233–244. North Holland Publ., Amsterdam.

Gelius, U., Heden, P. F., Hedman, J., Lindberg, B. J., Manne, R., Nordberg, R., Nordling, C., and Siegbahn, K. (1970). *Physica Scripta* **2,** 70.

Helmer, J., and Weichert, N. (1968). *Appl. Phys. Lett.* **13,** 266.

Hnatowich, D., Hudis, J., Perlman, M., and Ragaini, R. (1971). *J. Appl. Phys.* **42,** 4883.

Novokov, T., and Prins, R. (1972). "Electron Spectroscopy," pp. 821–826. North Holland Publ., Amsterdam.

Robinson, H. (1923). *Proc. Roy. Soc.* **A104,** 455.

Siegbahn, K. *et al.* (1967). "Electron Spectroscopy for Chemical Analysis." Almquist and Wiksells, Uppsala.

Siegbahn, K. *et al.* (1969). "ESCA Applied to Free Molecules." North Holland Publ., Amsterdam.

Siegbahn, K., Gelius, U., Siegbahn, H., and Olsen, E. (1970). *Phys. Lett.* **32A,** 221.

Steinhard *et al.* (1972). "Electron Spectroscopy," pp. 557–567. North Holland Publ., Amsterdam.

Wagner, C. D. (1972). "Electron Spectroscopy," pp. 861–884, North Holland Publ., Amsterdam.

Wertheim, G. K., and Rosencwaig, A. (1971). *Phys. Rev. Lett.* **26,** 1179.

Author Index

Numbers in italics refer to the pages on which the complete references are listed.

D

E

F

G

H

Mc

M

N

O

P

R

S

T

U

V

W

Y

Z

Subject Index

A

B

C

D

S

T

U

V

X

Z